Applying Lean Six Sigma
in the Pharmaceutical Industry

This book is dedicated to my dear friend Cordell
whose support has always been behind my success in this business

and to my wife Debbie
whose spirit has always been behind my success in life.

Applying Lean Six Sigma in the Pharmaceutical Industry

BIKASH CHATTERJEE

Routledge
Taylor & Francis Group

LONDON AND NEW YORK

First published in paperback 2024

First published 2014 by Gower Publishing

Published 2016 by Routledge
4 Park Square, Milton Park, Abingdon, Oxon OX14 4RN

and by Routledge
605 Third Avenue, New York, NY 10158

Routledge is an imprint of the Taylor & Francis Group, an informa business

British Library Cataloguing in Publication Data
A catalogue record for this book is available from the British Library.

Library of Congress Cataloging-in-Publication Data
Chatterjee, Bikash, author.
 Applying lean six sigma in the pharmaceutical industry / by Bikash Chatterjee.
  p. ; cm.
 Includes bibliographical references and index.
 ISBN 978-0-566-09204-6 (hardback)
 I. Title.
 [DNLM: 1. Technology, Pharmaceutical--standards. 2. Drug Discovery--organization & administration. 3. Efficiency, Organizational. 4. Technology, Pharmaceutical--organization & administration. 5. Total Quality Management--methods. QV 778] RS100

 338.4'76151--dc23
                                                                          2013042257

 ISBN: 978-0-566-09204-6 (hbk)
 ISBN: 978-1-03-283837-3 (pbk)
 ISBN: 978-1-315-56741-9 (ebk)

 DOI: 10.4324/9781315567419

Contents

List of Figures

List of Tables

List of Abbreviations

ANDA	Abbreviated New Drug Application
API	Active Pharmaceutical Ingredient
APR	Annual Product Review
AQL	Acceptable Quality Level
BLA	Biologic Licence Application
BRIC	Brazil, Russia, India, China
BVA	Business Value Added Activity
CAPA	Corrective and Preventative Action
CDER	Center for Drug Evaluation and Radiology
CE	Conformite Europeene
cGMP	Current Good Manufacturing Practice
cGxP	Current Good X Practice
CMA	Critical Material Attributes
CMC	Chemistry and Manufacturing Controls
COA	Certificate of Analysis
COC	Certificate of Compliance
CPP	Critical Process Parameters
CQA	Critical to Quality Attributes
CRO	Contract Research Organization
cVSM	Current Value Stream Map
DCOV	Define, Characterize, Optimize, Verify
DFLSS	Design for Lean Six Sigma
DMADV	Define, Measure, Analyze, Design, Verify
DMAIC	Define, Measure, Analyze, Improve, Control
DOE	Design of Experiments
ERP	Enterprise Resource Planning
eVSM	Electronic Value Stream Map
FDA	Food and Drug Administration
FFA	Force Field Analysis
FIFO	First-In-First-Out
FMEA	Failure Modes and Effects Analysis
FTA	Fault Tree Analysis
FTE	Full-time Employee
fVSM	Future Value Stream Map

G&O	Goals and Objectives
GE	General Electric
GMP	Good Manufacturing Practices
GRR	Gauge Repeatability and Reproducibility
HACCP	Hazard Analysis and Critical Control Points
HAZOP	Hazard and Operability Method
HPLC	High-Performance Liquid Chromatography
ICH	International Committee on Harmonization
IDOV	Identify, Design, Optimize, Verify
IND	Investigational New Drug
IOQ	Installation Qualification and Operational Qualification
IQA	Incoming Quality Assurance Inspections and Testing
ISPE	International Society for Pharmaceutical Engineering
IVIVC	In-Vitro In-Vivo Correlation
KPIV	Key Process Input Variables
LIMS	Laboratory Management Information System
LSS	Lean and Six Sigma
LTPD	Lot Tolerance Percent Defective
MBR	Manufacturing Batch Record
MS	Mass Spectroscopy
NDA	New Drug Application
NOR	Normal Operating Range
NVA	Non-Value Added Activity
OEE	Overall Equipment Effectiveness
OMED	One Minute Exchange of Dies
OOS	Out of Specification
OpEx	Operational Excellence
PAR	Proven Acceptable Range
PAS	Prior Approval Submission
PAT	Process Analytical Technology
PCE	Process Cycle Efficiency
PD	Pharmacodynamic
PDA	Parenteral Drug Association
PDCA	Plan–Do–Check–Act
PDQC	Product Development Quality Control Laboratory
PDUFA	Prescription Drug User Fee Act
PK	Pharmacokinetic
PLC	Programmable Logic Controllers
PMA	Pharmaceutical Manufacturers Association
PPS	Pilot Plant Services
PQAS	Pharmaceutical Quality Assessment System

QbD	Quality-by-Design
QMS	Quality Management System
R&D	Research and Development
RCE	Rapid Changeover Exercise
RPN	Risk Priority Number
RTD	Resistance Temperature Detector
SCADA	Supervisory Control and Data Acquisition Systems
SME	Subject Matter Expert
SMED	Single Minute Exchange of Dies
SOP	Standard Operating Procedure
SPC	Statistical Process Control
TPDS	Toyota Product Development System
TPS	Toyota Production System
TTM	Time To Market
USP	United States Pharmacopeia
VA	Value Added Activity
VOC	Voice of the Customer
VSM	Value Stream Map

Preface

In theory, there is no difference between theory and practice.
But, in practice, there is.

Yogi Bara

When I was approached to write this book my first thought was, "Do we really need another book on Lean and Six Sigma?" The principles of continuous improvement are certainly not new. From Shewart and Deming to Ohno and Shingo the benefits of Lean Manufacturing and Six Sigma are well recognized. However, despite the well-publicized benefits the pharmaceutical and biotech industries have been slow to fully embrace these principles and benefits. Don't get me wrong, our industry was as enamored with the potential for Lean and Six Sigma as any industry was when Jack Welch first espoused the virtues of building and running an organization through Six Sigma. The body of evidence has continued to grow as to the benefits of the methodology as companies such as Southwest Airlines, Bank of America, Pfizer and many others have proclaimed the virtues of its structured methodology and tangible business improvement entitlements. However, the level of organizational inertia required to fully embrace Six Sigma to the same levels as General Electric (GE) require significant corporate commitment.

Lean Manufacturing, on the other hand, required very little organizational inertia to realize efficiency improvements. Lean's simpler concepts had the potential for rapid process improvement without having to invest in extensive education in the methodology. However, the luster has diminished somewhat because of quality issues over the last decade. Given the other market challenges, our industry has lost its enthusiasm for Lean and Six Sigma (LSS).

However, as a Lean Six Sigma Master Black Belt, the benefits and principles of these methodologies have been firmly demonstrated. Throughout my career I have seen countless case studies describing the deployment of the Define, Measure, Analyze, Improve, Control (DMAIC) method only to stumble at the project's end. I believe what sets successful Lean corporate cultural transformations apart from the rest revolves around an integration of LSS philosophies into the drug development life cycle. Lean professionals that insist on dealing with processes without fully understanding the role within the business's strategic performance miss the opportunity to connect the dots.

Features of this Book

This book is designed to provide a glimpse into how the theory behind Lean Manufacturing and Six Sigma relates to the pharmaceutical environment. The assumption is the reader is familiar with the basic principles and tools of both Lean and Six Sigma but perhaps has a limited understanding of the drug development process and environment and its relationship to LSS improvement initiatives. This book emphasizes the need to never lose sight of the compliance and regulatory consequences of improvement. I have a saying that I always reinforce with every LSS initiative: "Better is not always good, it's just different." I use this motto to emphasize that we cannot turn a blind eye to process and product understanding when improving processes. This book will take you through several of the major work centers of the drug development process. In some cases the case studies are focused on small molecule pharmaceuticals but the principles and approaches applied can be equally applied to biotech and medical device processes.

How Can this Book Help?

This book will take the theory of Lean Manufacturing and Six Sigma and put it into action. Each chapter includes discussion around the principles to consider for the application of LSS principles and tools and culminates in a case study application for that application. The book is organized to reflect the major work centers involved in the drug development life cycle. Each chapter is standalone but together they drive home the necessary synergy between LSS and compliance sensibilities required to be successful in our industry.

I start each chapter with a quote as a preamble to reinforce the point of each chapter and provide a subtle reminder that the effective transformation to a LSS culture is more than the thoughtful application of tools and data; it is the realization of a new way of thinking about processes and their relationship to business performance.

Chapter 1 is a primer on the basic elements of the drug development process. Understanding the migration from molecule discovery through clinical manufacturing, scale-up and commercial manufacturing is essential to correctly applying the tools of LSS.

Chapter 2 delves into the tools and principles surrounding risk management. Risk management has become a central component to the drug development process from product design to process and facility design. It is a central component to every process design activity, especially those that have a compliance element. Understanding the risk management tools, their strengths and application is an essential component of any improvement discussion in today's pharmaceutical organization. I end this chapter with a discussion of the emerging markets since there are very few organizations in the US that do not have an overseas component to them. Recognizing the challenges created by disparate business practices, cultural and communication differences drives home the value of applying risk management principles to any improvement initiative.

Chapter 3 starts at the beginning of the drug development process and looks at the application of Lean in the product development environment. The chapter puts forth a model and structure that has been successfully deployed in several pharmaceutical organizations but is by no means the only approach that can be successful. It is more important to understand the elements of the model than the model itself. The one thing that the LSS practitioner must keep in mind is that the product development business model is evolving continuously, with many companies outsourcing their discovery and screening process offshore. However the basic tenets of knowledge management are still essential to reducing the risk of failure, time commitment and uncertainty of this critical early stage development.

Chapter 4 looks at the implementation of Lean Manufacturing in the laboratory and discusses the similarities between the laboratory environment and the shop floor. The case study is based upon applying the principles of LSS to a product development laboratory.

Chapter 5 discusses the considerations in applying LSS in a Pilot Plant environment. The Pilot Plant is a challenging and critical work center in the drug development life cycle because of its impact on both product design and process design and scale-up. A poorly operated Pilot Plant can have a profound impact on the performance of a business unit and affect its ability to move through its development pipeline. This case study is the only case study which is based upon a biotech process. I chose this because of the complexity of the work center environment and to illustrate the many facets of the operation to be considered when optimizing its performance.

Chapter 6 moves to the shop floor and discusses a manufacturing process for a transdermal patch. The case study illustrates how to apply several Lean tools to dramatically improve equipment productivity. More importantly the case study will illustrate the need to always consider the business need when focusing on improvement. This goes beyond just the pure entitlement of the burning platform and includes the risk benefit considerations that must be made regarding strategic, legal and compliance risk.

Chapter 7 moves beyond routine manufacturing and discusses the considerations for deploying LSS when considering a Process Analytical Technology (PAT) project. The chapter puts forth a practical model and describes the considerations for realizing the benefits of PAT.

My hope is this book will be the missing piece to your Lean Manufacturing and Six Sigma library and will provide some insight into the qualities and considerations that are necessary to be successful in deploying LSS in the pharmaceutical environment.

Chapter 1
The Product Lifecycle and Quality Philosophy

Learning is more than absorbing facts, it is acquiring understanding.

William Arthur Ward

Background

There is an old saying that says: "To know your future your must know your past." You may ask what has this to do with Lean Manufacturing and Operational Excellence (OpEx). The truth is any improvement effort must understand the factors that formed its current state if an effective improvement is to be achieved. Without this understanding it is easy to overlook the constraints that can interfere with change.

Over the last three decades the industry has undergone tremendous change. The drivers for change have been varied. Certainly the Food and Drug Administration (FDA) has played a significant role in how we develop products and how we define product quality. These drivers have not been without heartache. In the 1980s, much of ethical pharma was under scrutiny as the FDA stepped up enforcement of its regulatory guidelines. Water systems and computerized systems were targeted in an effort to push the industry toward what the FDA felt was a higher standard of assurance. Water systems in particular were the first specific system which articulated design criteria that must be present in order to support a claim that the system was validated. Little did we know the seeds of Quality-by-Design (QbD) were being introduced to us in the guise of these new guidance documents.

Computerized systems were the next hurdle as the industry attempted to take advantage of the rapid evolution of PCs and automation technology. These improvements challenged us to examine closely the value of our Quality Assurance activities in terms of truly establishing product quality. The FDA themselves were forced to reconsider their position on several key requirements. Software code reviews came into the spotlight when the agency suspended the launch on Upjohn's highly automated manufacturing process for its Minoxidil product line because a 100 percent code review had not been completed as part of the validation exercise. Faced with hundreds of thousands of lines of code it was not realistic to do a 100 percent inspection of the code, which would take years, nor was it likely to be effective in ensuring good code development and structure. In lieu of this the agency and industry began discussions surrounding what could be effective in order to ensure final product quality. The FDA recognized that there was a philosophical divergence between their approach of oversight as a foundation for quality and industry's desire to improve through technology with the issuance of 21 CFR 210 Part 11 guidance for electronic signature and data retention. The guidance fundamentally encouraged companies to not modernize because of the prohibitively costly and complex regulatory requirements. The agency recognized this and publicly committed to developing a new guidance which would be more consistent with demonstrating process robustness, however this has not yet been issued.

It is reasonable to ask what makes the pharmaceutical industry so different from other industries, such as the semi-conductor or automotive industry, when it comes to implementing Lean Manufacturing and Six Sigma. I believe the difference lies in how we define and measure quality as part of the drug development process. As a continuous improvement practitioner it is easier to do more harm than good if the compliance commitments are not fully understood before attacking the problem. So let's explore the steps required to develop drug products and demonstrate they are safe and efficacious.

Product Development Lifecycle

When talking about drug development it is important to realize that the basic premise behind identifying and developing new drug therapies has not really changed since the 1950s but the level of scientific rigor required to navigate this development roadmap has changed dramatically. The basic elements of drug development are shown in Figure 1.1.

Drug Discovery

As with all industries, the product design process represents a key opportunity for improving business performance. The drug development process is the equivalent of the product design process in other industries. The first step

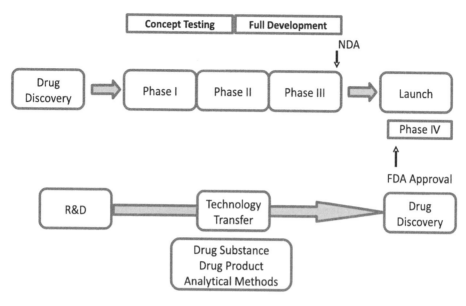

Figure 1.1   Product development lifecycle

in the process is drug discovery. Drug discovery begins with screening a large number of synthetic and naturally derived compounds for desirable characteristics. This activity is complicated by the fact that there are several routes to establishing what could be a promising compound. One approach is to take a lead compound and modify its molecular structure in an effort to make it more effective or reduce its side effects. Alternatively, some compounds are selected because they provide a potential advantage in terms of the disease's mechanism of action. Obviously, if a more efficient approach to molecule screening could be devised, the potential for reducing the development cost and time to market could be significant. Unlike other industries, product cost and marketing characteristics rarely creep into this phase of the lifecycle. Improving this process requires stepping out of the classical business performance mindset and devising a more effective approach to harnessing scientific inquiry.

In addition to compound identification there are other critical studies to be performed during this pre-clinical phase before putting a molecule into humans. Animal models are selected in an effort to detect overt drug toxicity. Toxicity studies attempt to assess the carcinogenicity, reproductive toxicity and teratogenic potential of the new molecule. Studies are also conducted in order to estimate the dose-response of pharmacological and toxic effects. These studies provide the first evidence of clinically relevant activity.

PHASE I

Given a promising target molecule, an Investigational New Drug (IND) application is filed requesting permission from the FDA, based upon the pre-clinical activities to test the molecule in humans. Basic drug therapy formulation studies have begun at this stage with a focus on achieving physical therapeutic performance requirements. Phase 1 clinical trials have several objectives but the primary endpoint is assessing the drug's safety. Classically, Phase 1 clinical trials for small molecules are conducted in healthy volunteers. The studies are open label, dose escalation studies designed to determine the maximum tolerated dose and the optimum biologic dose. In addition, baseline pharmacokinetic (PK) and pharmacodynamic (PD) measurements are taken. These studies often give the initial indication of the drug's potential effectiveness. For many small molecules the first discussions around establishing an In-Vitro In-Vivo Correlation (IVIVC) begin.

PHASE 2

Phase 2 clinical trials focus on evaluating the drug's performance in a larger number of the target population. These are hypothesis-generating studies and expand upon the safety and efficacy dosing or dosing regimen evaluated in the Phase 1 studies. These studies explore potential endpoints, study population and disease characteristics that can be used for the design of the Phase 3 clinical studies. These studies are generally randomized and blinded and require rigid adherence to the clinical protocol. Often Phase 2 studies include some sort of dose-ranging component in the target population. This data can be very important downstream when, conducting the risk management analysis of the process and final product, they may or may not include a control for clinical comparison. The data from these studies are intended to be supportive of the Phase 3 data generated in the next phase of development.

PHASE 3

Phase 3 clinical studies represent the final clinical step in establishing a drug's therapeutic effectiveness. These are large-scale multi-site trials in the intended patient population. These studies are usually randomized, double-blinded studies and are often placebo controlled. The endpoints of these studies are chosen to demonstrate maximum clinical benefit. These are the pivotal studies intended to support the New Drug Application (NDA) filing and market authorization. The data from these studies will generate the information required for the product labeling and risk–benefit information for the intended target population. The assumption within most regulatory organizations is that the key quality management systems (QMS) and final manufacturing processes are in place for the process in order to ensure the results of the clinical trial will be representative of the final commercial process and product. The reality in many small molecule processes in pharma and most biotech processes is that this is not the case and some sort of comparability argument will be required in order to move to high-volume commercial operations. How this is done effectively will be discussed in detail in Chapter 3 which discusses the considerations in applying Six Sigma in the development environment.

PHASE 4

Phase 4 studies are conducted post-approval. The intent of these studies can be quite diverse. Often they are commitments made to the FDA during the approval process intended to gather additional safety information in the intended target population or to gather information on performance in specific populations, or using different or extended dosing regiments. Many times, for example, a drug will be approved for adult use and a Phase 4 study will be conducted in children in order to gain approval for this expanded market. Phase 4 studies could also be used to measure performance against a competitive product in order to put forth a clinical case for superiority in the marketplace.

As the product moves through these escalating levels of clinical scrutiny the organization's QMS and manufacturing systems must escalate as well. An escalating model offers the potential for quickly realized improvements in efficiency and performance, especially early in the drug's development lifecycle, but must be done thoughtfully in order to not compromise the compliance commitment of a regulated drug.

The Philosophy Behind Quality

The history of quality and quality management infrastructure mimics that of other industries. Evidence of the key principles from Deming, Juran, and Crosby can be found in most pharmaceutical quality systems today.

Each of these quality gurus approached the business of quality from a slightly different perspective. All recognized the need to create an organizational environment which was conducive to understanding the process and its relationship to measurement and performance. The role of management in fostering and creating this environment was central to establishing an effective

QMS. While they agreed with this basic principle they disagreed on the role each played in achieving this environment. Deming felt it was important for managers to eliminate fear amongst the organization to promote the awareness and rapid resolution of quality issues. Juran, however, felt a certain amount of fear was good for a quality organization in terms of promoting execution and follow through. Crosby remained vague in terms of the merits of either approach. All three agreed, however, that quality must always be of strategic importance, right through manufacturing and sales in order to be successful in the marketplace.

Pharma has built their reputation on this principle but, in practice, the variability associated with interpreting these basic principles is enormous across the industry. When considering Lean and Six Sigma (LSS) initiatives within a pharmaceutical organization it is helpful to understand the basic principles each of these quality architects espoused before considering the scope and success metrics for a project. While productivity and revenue generation may resonate with the senior management team, it is typically the furthest consideration from a quality professional's mind. I am not saying that quality professionals don't care about making money for the business but, rather, they are in the business of risk management. In this case risk is both the risk to the customer and to the producer. A failure in either case can have a profound impact on business performance.

Deming's 14 Points

Deming believed it was management's job to create the right quality philosophy and architecture necessary to support the business. He did not believe this meant being strict and conservative all the time in terms of defect deterrence and detection, rather doing the appropriate level of Quality Assurance for the task before them.

The 14 points are a basis for transformation of American industry. Such a system formed the basis for lessons for top management in Japan in 1950 and in subsequent years. The 14 points apply anywhere, to small organizations as well as to large ones, to the service industry as well as to manufacturing. They apply to any division within a company.

Deming's 14 points are summarized as follows:

1. Create constancy of purpose toward improvement of product and service, with the aim to become competitive and to stay in business, and to provide jobs.

2. Adopt the new philosophy. We are in a new economic age. Western management must awaken to the challenge, must learn their responsibilities, and take on leadership for change.

3. Cease dependence on inspection to achieve quality. Eliminate the need for inspection on a mass basis by building quality into the product in the first place.

4. End the practice of awarding business on the basis of price tag. Instead, minimize total cost. Move toward a single supplier for any one item, on a long-term relationship of loyalty and trust.

5. Improve constantly and forever the system of production and service, to improve quality and productivity, and thus constantly decrease costs.

6. Institute training on the job.

7. Institute leadership. The aim of supervision should be to help people and machines and gadgets to do a better job. Supervision of management is in need of an overhaul, as well as supervision of production workers.

8. Drive out fear, so that everyone may work effectively for the company.

9. Break down barriers between departments. People in research, design, sales, and production must work as a team, to foresee problems of production and in use that may be encountered with the product or service.

10. Eliminate slogans, exhortations, and targets for the workforce asking for zero defects and new levels of productivity. Such exhortations only create adversarial relationships, as the bulk of the causes of low quality and low productivity belong to the system and thus lie beyond the power of the work force.

11. a. Eliminate work standards (quotas) on the factory floor, substitute leadership.

b.   Eliminate management by objective. Eliminate management by numbers and numerical goals, substitute leadership.

12.   a.   Remove barriers that rob the hourly paid worker of his right to pride in workmanship. The responsibility of supervisors must be changed from sheer numbers to quality.

b.   Remove barriers that rob people in management and engineering of their right to pride in workmanship. This means, *inter alia*, abolishment of the annual or merit rating and management by objective.

13.   Institute a vigorous program of education and self-improvement.

14.   Put everybody in the company to work to accomplish the transformation. The transformation is everybody's job.

When considering an improvement initiative in the pharmaceutical environment it is the QMS that represents the industry's key compliance element when contemplating Lean or Six Sigma initiatives. The QMS is the framework used to demonstrate that the drugs manufactured are safe and efficacious. The QMS framework has been defined in the September 2006 Guidance by FDA *Guidance for Industry Quality Systems Approach to Pharmaceutical CGMP Regulations*. There are four basic elements to the QMS: Management Responsibilities, Resource Management, Manufacturing/Lifecycle Requirements, and On-Going Evaluation Activities. These four elements supported by the quality organization's infrastructure must be designed to work together as a system in order to demonstrate a drug product has been made properly. These four elements are shown in Figure 1.2 as a closed loop system.

Management Responsibility

The role of management within the design and administration of the QMS is central to its effectiveness. The expected role of management within a well-defined QMS is defined in the September 2006 FDA guidance: *Guidance for Industry Quality Systems Approach to Pharmaceutical CGMP Regulations*. In it the guidance defines management as having six defined responsibilities within a well-designed QMS:

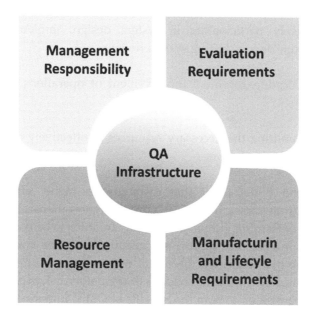

Figure 1.2   Pharmaceutical QMS framework

1.   provide leadership;

2.   structure the organization;

3.   build the quality system to meet requirements;

4.   establish policies, objectives, and plans;

5.   review the systems.

Provide Leadership

Organizational leadership is arguably the most essential component to an effective QMS. The overall QMS must be aligned with a manufacturer's strategic plans to ensure that the system is part of the manufacturer's mission and quality strategies. The quality management organization must stand on equal footing with other departments within the business. The QMS must show evidence that all levels of management provide support for the quality system by:

- actively participating in system design, implementation, and monitoring, including system review;

- advocating continual improvement of operations of the quality system;

- committing the necessary resources to effectively administer and execute the system.

It is the second bullet point that distinguishes highly effective quality organizations from classical pharmaceutical organizations. This speaks to the key points of Deming, Juran, and Crosby. In order to be comfortable with continuous improvement a quality organization must be comfortable with change. Change has historically been met with scrutiny within our industry and has been burdened with the shackles of regulatory risk exposure. In order to allow an organization to reap the benefits of OpEx, initiatives such as Lean Manufacturing and Six Sigma change must be embraced. Management's willingness to embrace this philosophy ties back to an effective risk management program and requires the Six Sigma or Lean Manufacturing leader to articulate the proposed changes within the context of compliance, regulatory and business risk.

Structure of the Organization

This requirement focuses upon the specific roles and responsibilities of the quality organization. The proposed QMS structure needs to be compatible with the organization's ability to generate quality products. Quality products can be defined as products that behave predictably and are generated and released from processes that are well characterized and understood. Understanding the quality architecture is essential to effectively proceeding with any form of improvement initiative. If the documentation and milestones are well understood from the outset then the disruption to the improvement activity can be greatly minimized. For example, if the OpEx team is looking at improving mixing efficiency within a compounding operation and is contemplating an alternative impeller design, understanding the QMS impact from the planned experimental exercise can save the team and organization a great deal of heartache. Questions that could be asked include: Is the current system qualified for Current Good Manufacturing Practices (cGMP) operation? If so, will this modification require a requalification exercise? Does a deviation need to be generated to take the system out of production? Does Quality Assurance need to approve the investigational protocol and final report? Are there limits filed in the Abbreviated New Drug Application (ANDA), New Drug Application (NDA), or Biologic License Application (BLA) that must be adhered to before moving ahead? Is it possible this experiment could impugn, or infer that the current process is somehow sub-standard or out of control? Anticipating the QMS risk and documentation requirements is essential to implementing OpEx in the pharmaceutical and biotech environment.

Build the Quality System to Meet Requirements

This requirement constitutes the primary challenge facing the industry today. Historically, the industry's QMS has been built upon a framework of inspection, testing, and documentation. Auditability has always been equated with transparency and in turn process control. We are learning today this is just one piece of the puzzle. The FDA states the following elements must be present in a well-defined quality organization structure:

- the scope of the quality system including outsourcing;

- the quality standards to be followed;

- the manufacturer's policies to implement the quality systems criteria and the supporting objectives;

- the procedures needed to establish and maintain the quality system.

Specifically the guidance highlights the need for a structured change control procedure along with a structured document control procedure to ensure critical records are captured, stored, and retrievable. Understanding these systems is a critical precursor to contemplating any form of OpEx activity. We will discuss this in detail in Chapter 2 in terms of the role this understanding plays in catalyzing organizational change.

Establish Policies, Objectives, and Plans

Policies, objectives, and plans under a modern quality system provide the means by which senior managers articulate their vision of and commitment to quality to all levels of the organization. The guidance specifically states: "Under a quality system, senior management should incorporate a strong commitment to quality into the organizational mission. Senior managers should develop an organizational quality policy that aligns with this mission; commit to meeting requirements and improving the quality system; and propose objectives to fulfill the quality policy."

OpEx initiatives must frame their improvement initiatives with these policies and procedures in mind. Short of this it is highly likely the improvement initiative may fail or do more harm than good, generating deviations for non-compliance with existing procedures and policies. It is important to remember that most of the industry in the US has gone through several cathartic changes in terms of its quality philosophy, usually as a result of heightened FDA enforcement. In the 1980s, nearly 80 percent of all ethical pharmaceutical companies were under consent decree for cGMP violations and breakdowns in quality systems. With each one of these enforcement actions the industry changed its baseline or benchmark for industry best practice. In many cases the response was to add additional compliance overhead to the process in an effort to tighten up control. These legacy policies and procedures today may seem ill conceived or in some cases not applicable, but unless they have been made obsolete or replaced with more current systems they must be adhered to.

Review the System

The final step in the process is to ensure the QMS structure, policies, and procedures meet the needs of the business and are in harmony with the latest regulatory requirements for the industry. This is a formalized process that should at a minimum consider:

- the appropriateness of the quality policy and objectives;

- the results of audits and other assessments;

- customer feedback, including complaints;

- the analysis of data trending results;

- the status of actions to prevent a potential problem or a recurrence;

- any follow-up actions from previous management reviews;

- any changes in business practices or environment that may affect the quality system (such as the volume or type of operations);

- product characteristics meeting the customer's needs.

While this process may not directly affect the OpEx initiative it represents a potential framework for change which, if harnessed correctly, can be used by the OpEx team to implement changes to existing systems.

In order to illustrate how these elements can impact an improvement initiative, let's review their role in an actual case study.

Case Study—Management Responsibility

We were asked to assist a small molecule drug manufacturer in solving a process problem with a capsule product that had been on the market for nearly ten years. Historically, approximately 10 percent of the lots had failed its dissolution specification on an annual basis. Eventually the process had drifted to a point where it was not possible to make product that could meet specification. A small project team was convened in order to correct the problem and return the product back to commercial manufacturing. The team used a structured brainstorming approach to gather data.

Starting with the development report, the team evaluated the product design in order to try and understand what the possible mechanism problems could be. The team noted that there was no In-Vitro In-Vivo Correlation (IVIVC) established for the product, which meant the team could make minor modifications to the drug release profile which would most likely have limited impact on the drug's therapeutic performance. This was an important realization prior to experimentation and allowed the team to explore the process space with some level of immunity. A historical review of the product's performance and process capability was made and used to specifically establish success metrics for the improvement initiative. The team used a Kepner Tregoe format for evaluating the data against the potential failure modes. This drove the team to identify a short list of possible investigations which could be used to understand and demonstrate the root cause of the process failure. The results of this analysis were communicated to both the technical leadership along with compliance and regulatory leadership for discussion. This was done at the outset for two purposes. First to ensure all the key stakeholders understood the scientific assessment and their consequences and second to provide an opportunity for the organization to highlight potential obstacles in moving forward with corrective actions.

The discussion was critical since it identified several changes which would require regulatory submissions which, for this product, would not be the desired path forward. It also highlighted several internal procedures which had been created as a result of past regulatory actions which would significantly lengthen the time to any process improvement but were mandatory based upon their past commitments to the FDA. With these new requirements in mind the team proceeded to conduct its root cause investigation. Without this information the team would have pursued solutions which may have been effective but in conflict with the compliance and regulatory strategy of the site.

RESOURCE MANAGEMENT

Resource management may seem obvious to most when considering the execution of a well-defined QMS. However, understanding how resources are deployed can become critical to not only assembling the right members of your OpEx team but also for executing your improvement plan. The FDA guidance defines resource management as having four defined responsibilities within a well-designed QMS:

1. general arrangements;

2. personnel development;

3. facilities and equipment;

4. control of outsourced operations.

General arrangements

Ironically this is an integral component to presenting a defensible QMS and is a problem faced by many start-up ventures that are focused on product development. The structure and resources must be sufficient to convey that the organization is capable of effectively administering the QMS and operate in a cGMP manner. At a minimum, senior management is responsible for demonstrating there are adequate resources:

- to supply and maintain the appropriate facilities and equipment to consistently manufacture a quality product;

- to acquire and receive materials that are suitable for their intended purpose;

- for processing the materials to produce the finished drug product;

- for laboratory analysis of the finished drug product, including collection, storage, and examination of in-process, stability, and reserve samples.

Personnel development

Personnel development pertains largely to ensuring there is an embedded program for training within the QMS. At a minimum quality managers are responsible for:

- evaluation of training needs;

- provision of training to satisfy these needs;

- evaluation of effectiveness of training;

- documentation of training and/or re-training.

Understanding the training infrastructure can be a critical element to moving forward with a planned improvement or corrective action. If not considered, the organization's ability to implement and realize the benefits of the improvement initiative can be significantly delayed.

Facilities and equipment

The quality unit is responsible for reviewing all facility and equipment design criteria against the basic tenets of cGMP and the organization QMS. This includes all preventive maintenance systems and calibration systems which apply to the facility and equipment. This is an often overlooked component to many OpEx initiatives. Understanding the procedures used for calibration and duty cycles used for maintenance that have been established for equipment and critical utilities can be an essential part of understanding the sources of variation in a process. If equipment stability cannot be established at the outset of an investigation it is nearly impossible to establish process predictability.

Control of outsourced operations

As most business strategies look to reduce overhead and standard costs, most include some element of outsourcing. Wherever the outsourcing takes place within the product development lifecycle, the primary manufacturer's quality unit is still responsible for release and disposition of the final commercial product. In order to support an outsourcing strategy most QMSs include a defined supplier qualification program which consists of a combination of a qualification audit, and contractual obligations defined within a quality agreement and supply agreement. Within OpEx the challenge lies in understanding the true process capability of the outsourced operation versus the pre-defined specifications. Many contract manufacturers and packagers support multiple markets, some of which may not require the procedural rigor of a pharmaceutical cGMP operation. In this case inspection, testing and auditability may not be enough to ensure process predictability and may require the manufacturer to carry out process characterization studies at the contract manufacturer in order to understand and control the basic process variation.

Case Study—Resource Management

A small molecule pharmaceutical manufacturer initiated a root cause exercise as part of a Corrective and Preventative Action program (CAPA) opened for a controlled release cardiovascular capsule that had been on the market for nearly ten years. A historical review of the product's performance indicated the process stability had been poor from day one. The team adopted a Six Sigma type methodology for the investigation, mimicking the Define, Measure, Analyze, Improve and Control (DMAIC) roadmap. Based upon the review, several key facts were established. First, the product had exhibited a high variability since its commercial introduction and possessed a process capability well below one, for example, a much less than three sigma process. The product formulation contained several components which could be sensitive to environmental conditions. Once again the focus of the CAPA was a dissolution failure. Also, no IVIVC had been established for the product so the team operated with some immunity in that it could modify the release profile lightly without affecting the product's therapeutic behavior. The product is spray coated using a Wurster column process then dried in tray ovens. Since the product dissolution may be affected by temperature variation, the team focused on characterizing and stabilizing the ovens before beginning any experimentation. The stabilization exercise revealed several key factors which could have derailed the investigation. First the Installation Qualification and Operational Qualification (IOQ) protocol did not require verifying the airflow through the oven. Second, the current calibration procedure for the temperature control was single-point verification based upon a Resistance Temperature Dectector (RTD) in the inlet air stream. However, because the oven needed to be cleaned during product changeover, the RTD could be moved anywhere within the oven, making the actual oven control and temperature distribution quite variable. When the maintenance team was asked to calibrate the air flow controller they reported that the system had a very large offset from the controller setpoint. A deeper investigation revealed the procedure used by the technicians was incorrect and did not account for the duct geometry in calculating the F-factor. Temperature mapping the unit revealed large temperature variations between the top and bottom shelves of the dryer. A closer look revealed the gasketing for the oven doors was completely worn away, leaking air from the ovens and the gasket on the bottom of the oven was never installed. It is important to remember the ovens had a calibration procedure and sticker on the equipment. When the technicians were asked if the equipment was part of the PM program the answer was yes. In response to the observations by the team a new calibration procedure for the air flow was develop and formalized as a procedure. Technicians were then trained in this new procedure. In addition, new gasketing was installed on all of the ovens and the gaskets were identified as a critical element in the PM procedure. Finally the RTD location was specified in each oven and the control loop tuning for the proportional–integral–derivative (PID) controller repeated to minimize variation around the setpoint, reducing the total variability from ± 5°C to ± 1°C. Only at this point did the team begin its experimentation to quantify the contribution from temperature in terms of the product's performance.

MANUFACTURING AND LIFECYCLE REQUIREMENTS

There is considerable overlap between the QMS and operational systems within the pharmaceutical environment. The construct of this overlap is at the very heart of the challenge for the industry when trying to embrace new process development philosophies such as QbD. The FDA states that there are four basic components in terms of manufacturing and lifecycle requirements in a modern QMS. They are:

1.   design, develop, and document product and processes;

2.   examine inputs;

3.   perform and monitor operations;

4.   address non-conformities.

DESIGN, DEVELOP, AND DOCUMENT PRODUCT AND PROCESSES

The FDA guidance cites the need to ensure the product and process is defined from design through commercial introduction. Procedures must be in place to define responsibility for development, scale-up, technology transfer and commercial manufacturing along with all supporting operations such as quality control sampling, maintenance and calibration, and batch documentation. Specific documentation requirements defined in the guidance include:

- resources and facilities used;

- procedures to carry out the process;

- identification of the process owner who will maintain and update the process as needed;

- identification and control of important variables;

- quality control measures, necessary data collection, monitoring, and appropriate controls for the product and process;

- any validation activities, including operating ranges and acceptance criteria;

- effects on related process, functions, or personnel.

The process documentation is a critical consideration for any OpEx exercise. Often the batch record instructions represent the regulatory commitments for the product despite what is listed in the Chemistry and Manufacturing Controls (CMC) section of the filing. Understanding what is captured and how it is captured in the batch record can be an important consideration when evaluating process improvements. Especially for legacy products, where the original filing may not contain the scientific rigor modern filings contain, opening up the filing to the scrutiny of regulatory authorities may do more harm than good. Often by understanding what is and is not captured, a plausible path forward can be identified that balances regulatory risk with compliance and process stability considerations.

EXAMINE INPUTS

The FDA deems anything that goes into the product as an input to the process regardless of whether they are purchased by the manufacturer or made by the manufacturer. The expectation from a robust quality system is that all inputs to the manufacturing process are reliable because quality controls will have been established for the receipt, production, storage, and use of all inputs. Typical tools for achieving this level of reliability are establishing Incoming Quality Assurance Inspections and Testing (IQA), utilizing supplier audits in combination with a risk-based prioritization scheme to ensure critical suppliers and contract manufacturers provide input material that will perform as expected. Once a history of process stability has been established, input materials can move from an IQA approach to a Certificate of Analysis (COA) or Certificate of Compliance (COC), supplemented with periodic reassessments. The challenge with any OpEx initiative is recognizing the true criticality of these inputs to the overall process stability. Often the specifications reflect United States Pharmacopeia (USP) compendia limits or the supplier's commercial ranges, and are not in line with the final product's process requirements. Understanding both the criticality of the process input and the resolution of the measurement tool used to assess the input variability is the primary challenge facing an improvement initiative.

PERFORM AND MONITOR OPERATIONS

The goal of establishing, adhering to, measuring, and documenting specifications and process parameters is to objectively assess whether an operation is meeting its design and product performance objectives. In a robust quality system, production and process controls should be designed to ensure that the finished products have the identity, strength, quality, and purity they purport or are represented to possess. In order to do this effectively the critical variables which drive the product's design and performance should be well understood during the research and development (R&D) phase of the product's lifecycle. In a well-defined product development process the critical factors required for moving the product to commercial scale also become part of the design criteria in addition to meeting the product's therapeutic performance metrics. Moving the OpEx mindset into the development environment must be done thoughtfully because of the organizational mindset and charter which focuses more scientific inquiry than efficiency and effectiveness. Generic companies are an excellent example of the challenge in moving the OpEx sensibility into the development environment. Many generic company R&D groups are measured by their ability to be the first to file because of the market exclusivity granted to the first to file their ANDA. However, there is often no follow-up in terms of the quality and effectiveness of the filing or the cost and effectiveness of the commercial scale-up effort. Ensuring meaningful metrics are in place is the critical last step in ensuring a stable and predictable process remains in control.

ADDRESS NON-CONFORMITIES

Under a quality system, if a product or process does not meet requirements, it is essential to identify and/or segregate the product so that it is not distributed to the customer. The investigation, conclusion, and follow-up must be documented to ensure that a product conforms to requirements and expectations. It is important to measure the process and the product attributes (for example, specified control parameters, strength) as planned. Discrepancies may be detected during any stage of the process or during quality control activities. Not all discrepancies will result in product defects; however, it is important to document and handle discrepancies appropriately. Effective analysis and resolution of deviations and CAPA investigations is one of the Achilles heels of this industry. A strong quality system will track and monitor reoccurring deviations and evaluate, on a periodic basis, along with the effectiveness of the proposed corrective and preventive actions.

The challenge with evaluating most non-conformities is understanding the full implication of the deviation. Without a clear mechanistic understanding of the relationship between the input variability, process control, and specifications it is nearly impossible to fully anticipate the full consequence of a deviation. An intermittent failure may be an isolated event or an early indication of equipment or process drift. The challenge for OpEx initiatives is to compensate for this knowledge gap. This often requires starting with fundamentals by mapping the product's design criteria against the available data for the operational and quality systems. It is easy to see how, given the lack of clarity in most development and scale-up technical packages, a minor deviation can escalate to a complex investigation. As OpEx practitioners, our challenge is to apply a methodology for evaluation which can definitively separate the noise in the process from the true process variation in order to truly address the root cause of the non-conformity.

Case Study—Manufacturing and Lifecycle Requirements

A manufacturer of an aseptically manufactured transdermal patch was manufacturing its Phase 3 clinical supplies. The patch consisted of three different polymer materials which were brought together for production at the manufacturer on a custom-designed transdermal manufacturing line. Two of the materials were actually laminates which were converted by a third-party contract manufacturer before delivery to the manufacturer. The three layers were held together by an adhesive which was applied to one of the materials at the supplier. As this was an aseptic product, all three components were sterilized using gamma radiation. The sterilization process had been validated against the ISO 11137 VD_{max} method, meaning sterility was being established by verifying a minimum radiation dose of 25KGy. Since most polymers and adhesives do not like irradiation, the adhesive used in this product system was specially selected to withstand the rigors of irradiation. All three lots were manufactured for the Phase 3 clinical supplies and additional lots were built for the registration batches to demonstrate product stability and establish the product expiration dating. One month into the stability program the quality control lab began reporting degradation of products and impurities that had not been seen in previous lots. The clinical readiness activities were stopped at the Contract Research Organization (CRO) and an investigation was initiated. The organization employed a structured root cause analysis program which mimicked the DMAIC roadmap milestones.

The investigation concluded several failures within the QMS which led to the stability results. First the supplier that applied the adhesive as part of one of the laminates changed one of its sub-suppliers using a material that was substantially equivalent, in their words. Second, the change control agreement in place with the supplier did not specifically reference sub-suppliers and hence did not flag to the supplier that they needed to notify the manufacturer of the substitution. Third, the IQA specifications were not designed to detect chemical differences in the material, only physical differences before irradiation. The Quality Assurance group had not developed and implemented a quality agreement with the supplier so it was unclear as to whose responsibility this new discrepant material was. To make matters worse, the manufacturer had a partner for marketing and this deviation triggered a heightened level of oversight by the partner as they lost confidence in the manufacturer's ability to manage its quality system.

Finally the investigation team was now tasked with finding a new adhesive that could meet the same physical and chemical specifications of the adhesive that was no longer available. Given the nature of the failure, the team was also tasked with establishing tests and metrics at the supplier and upon IQA that would be able to discern chemical differences in the material. The result was a six-month delay to the development program.

EVALUATION REQUIREMENTS

The last component to an effective QMS is an embedded structure for evaluation and continuous improvement. The guidance cites the following six elements in a effective QMS:

- analyze data for trends;

- conduct internal audits;

- quality risk management;

- corrective actions;

- preventive actions;

- promote improvement.

Analyze data for trends

An effective QMS is based upon a framework of understanding, monitoring, and corrective action. One critical aspect of achieving process and product understanding is trending analysis of critical data. The regulatory framework for all products within the US requires an annual review of each commercial product. These Annual Product Reviews (APR) capture the critical information associated with changes to the process along with its behavior based upon the critical parameters identified in the submission. There are minor differences between the US European and API (ICH Q7A) regulatory guidance documents in terms of what must be included in an APR, however the purpose of each is the same, to ensure the product stability and performance has not changed. Typical APRs include trend analysis of complaints, stability, returned and salvaged products, manufacturing performance (yield/failed lots), deviations and CAPAs opened, change controls opened, and specification changes among others. Often the APR is used as a trigger to focus the QMS internal audit schedule for the coming year in order to ensure critical processes of interest are working as expected.

Conduct internal audits

Internal compliance assessments are a critical tool for ensuring the QMS is working as intended. As with other procedures, audit procedures should be developed and documented to ensure that the planned audit schedule takes into account the relative risks of the various quality system activities, the results of previous audits and corrective actions, and the need to audit the complete system. Procedures should describe how auditors are trained in objective evidence gathering, their responsibilities, and auditing procedures. Procedures should also define auditing activities such as the scope and methodology of the audit, selection of auditors, and audit conduct (audit plans, opening meetings, interviews, closing meetings, and reports). It is critical to maintain records of audit findings and assign responsibility for follow-up to prevent problems from recurring. Audit findings can be a useful tool when crafting OpEx initiatives. Lean projects in particular should pay close attention to the findings from internal and external audits as they will highlight both internal areas of concern as well as external regulatory concern. Lean initiatives that do not address these concerns at the outset have the potential for increasing efficiency while moving a process out of compliance.

Quality risk management

Quality risk management has become the hallmark of the modern QMS. ICH Q9 and Q10 speak directly to the process of risk management and its integration into a pharmaceutical QMS. The tools for risk management have been in place for many years and have even been used on occasion within the development lifecycle. Understanding the strengths and weaknesses of these tools is central to ensuring the appropriate risk management path forward. The most commonly used risk management tools are:

- Fault Tree Analysis (FTA);

- Hazard Analysis and Critical Control Points (HACCP);

- Hazard and Operability Method (HAZOP);

- Failure Modes and Effects Analysis (FMEA).

The application of these tools will be discussed further in Chapter 2.

Corrective action

Corrective action is a reactive tool for system improvement to ensure that significant problems do not recur. The corrective action is the first half of the CAPA exercise. A typical CAPA program consists of the following steps:

- **Identification**—clearly define the problem.

- **Evaluation**—appraise the magnitude and impact.

- **Investigation**—make a plan to research the problem.

- **Analysis**—perform a thorough assessment and identify the root cause.

- **Action plan**—create a list of required tasks.

- **Implementation**—execute the action plan.

- **Follow-up**—verify and assess the effectiveness.

The weakest link in most corrective action plans is the identification of the root cause for the non-conformity. Highly successful investigations follow a structured methodology for the identification of the root cause. Recurring non-conformities with recurring corrective actions are a common observation during many FDA surveillance audits. A common rule of thumb for measuring the effectiveness of the root cause investigation is to intentionally create the non-conformity based upon the mechanistic understanding derived from the root cause investigation.

Preventive action

Preventive action is the final step in the CAPA exercise. Based upon the results of the corrective action, a preventive action plan should be designed to prevent the non-conformity from occurring. In order to ensure the preventive action is effective the current non-conformance frequency should be used as a baseline for comparison.

Promote improvement

The final element within an effective QMS is that it should support a philosophy of continuous improvement. This is perhaps the single biggest change in the compliance philosophy for this industry in decades and the greatest ally of an OpEx initiative. However, improvement without consideration for compliance or regulatory history will always come under scrutiny.

Case Study—Evaluation Requirements

A small molecule, Schedule 3 drug is manufactured as a transdermal system. The product is manufactured by first compounding the Active Pharmaceutical Ingredient (API) with excipients and extruding the mixture onto a laminate web. The web is then converted, die cut in the web, then wound up on a dispensing reel and stored in a foil pouch with desiccant. The final product is manufactured in several strengths and for several countries. The lowest-strength patch is manufactured with a clear backing while the higher-strength patch has a flesh-colored backing. Final packaging for all strengths takes place on a single packaging line which separates the patches from the master roll then pouches them and places them in trays for staging to a final cartoner. Two inspectors are placed on the line to visually inspect the quality of each patch before it enters the pouching station. Quality Assurance trends the incidence of successful changeovers between runs, tracking primarily stray patches that are found under or within the machine after a cleaning and change over. Similarly, Quality Control pulls samples from the final trays after they have been pouched and does destructive testing by opening the pouch samples and inspecting the patches and pouches against the product release specifications. Over a series of months the incidence of strays found during line clearance by Quality Assurance has been increasing. Similarly Quality Control has discovered intermittent lots which have pouches that contain patches with either incomplete labeling and/or mixed strength patches in incorrect pouches.

A deviation was issued for each lot along with a CAPA to address this escalating problem. A detailed evaluation of the pouching process identified that, periodically, patches dislodge from the master web and get stuck in various parts of the equipment. In some cases the patches wrap themselves around guide and nip rollers, in other cases they adhere to the guarding on the equipment. This made it very difficult to identify clear patches during the line clearance and changeover process. The incomplete patches found in the final pouch were attributed to a final web transfer device on the equipment which routinely drifts in and out of control. As a result, occasionally patches which have been rejected by the equipment and/or operators do not get physically removed from the line.

Despite these basic design issues, the Quality Assurance response was to increase the sample size for inspection and determine if this was an isolated excursion during the run or a more frequent event. Despite the obvious intrinsic variability in the process, the CAPA attributed the failures to operator error and closed out the investigation with operator retraining and a physical isolation barrier which discourages communication between operators. Not surprisingly the same failure appeared later. This reoccurring failure was cited as a major failure of the QMS during a routine FDA surveillance audit, requiring the manufacturing to stop until the issue could be definitively resolved.

Conclusion

In order to be effective as a continuous change agent within a regulated environment it is essential to understand the commitments made by the organization in terms of its development, compliance, and regulatory process. Other, less regulated, industries allow improvement initiatives to distill the path forward based upon the application of classical Lean principles such as "Takt time" and "Value Stream Mapping," looking only at efficiency opportunities. Pharma and biotech, however, require a strong understanding of where the organization is in terms of its process and product development maturity and its QMS evolution and history. By embracing these basic systems, which make up the foundation of the business, it will be possible to realize efficiency improvement and variability reduction opportunities that will enhance the organization's compliance position while satisfying the overall business objectives.

Chapter 2

Trouble in Paradise
and Catalyzing Change

We took risks, we knew we took them; things have come out against us,
and therefore we have no cause for complaint.

Robert Falcon Scott 1868–1912,
English polar explorer in Scott's last expedition (1913)

As I have said before, in order for LSS to be successful, compliance considerations must be at the forefront of the project. The most effective programs utilize continuous change agents that are steeped in regulatory and compliance experience. Understanding the FDA's thinking is central to defending an improvement of corrective action to a regulated process. Industry's relationship with the FDA and its sister regulatory organizations around the world have been tenuous at best. For years perceived to be in conflict with industry, the FDA has struggled to establish a framework with industry that was not strictly punitive but fostered some facility for dialogue. One example of this is the plethora of guidance documents issued by the agency in an effort to provide direction and articulate what the FDA would like to see in every facet of the drug development business. While this makes sense in theory the reality is the drug development process is a highly complex undertaking complicated by a myriad of specific details which can dramatically impact compliance exposure and public risk. As a result, FDA Compliance Inspectors were required to be experts in a vast array of guidances in order for this approach to be effective in a moving industry to some common level of understanding. Practically this was not easy to implement. As a result, enforcement was uneven across industry, promoting confusion and lack of consensus amongst both inspectors and industry. In an effort to promote common understanding, the FDA chose to focus on specific areas within the industry through a combination of focused surveillance inspections and presentations at industry professional organizations, such as the Pharmaceutical Manufacturers Association (PMA), the International Society for Pharmaceutical Engineering (ISPE) and the

Parenteral Drug Association (PDA). In the 1980s the focus was on water system design and testing and Computerized System Validation. The FDA's decision to specifically articulate requirements was a bold departure from past regulatory philosophy guidance approaches and reflected the philosophy of the Head of the FDA's Center for Drug Evaluation and Radiology (CDER) division, Joe Famulare, who strongly believed that the FDA and industry needed to find a way to work together for the common public good. Focusing on water systems was a brilliant move on the agency's part since water was used in almost every facet of the drug manufacturing and testing process. The decision to move away from unidirectional systems equipped with point-of-use filters for microbiological protection and mandate recirculating systems which could demonstrate escalating level of water quality improvement hit industry in the pocketbook, getting their attention and spurring dialogue.

The focus on computerized systems was an attempt to embrace the reality that automation and information management systems were becoming a larger and larger part of the drug development process. The integration of Programmable Logic Controllers (PLC), Vision Systems and Supervisory Control and Data Acquisition Systems (SCADA) offered industry the ability to easily gather, analyze, and adjust the process based upon real-time data. Analytical equipment was capable of directly capturing information from the equipment and storing it in a single database allowing Quality Assurance organizations to analyze stability data in many different ways. The FDA quickly recognized that their old philosophy of inspection as a paradigm for control would not work for computerized systems. With a compliance philosophy built around end product testing and documentation, audit trail computerized systems represented a curious conundrum. Another facet of adopting specific systems as platforms for introducing a new approach to compliance is that the FDA started to take on the persona of the inspectors that educated at industry organizations and were involved in enforcement of the FDA's new guidances. Hank Avalone, from the FDA's Atlanta district, became the de facto authority of water system design and qualification. Phil Piasecki, from the FDA's Philadelphia district, became the voice of computerized system validation, particularly when he shut down all Red Cross Blood Banks because they could not produce a compliant computerized system validation program. Similarly, Paul Motise became the object of industry's ire with his steadfast adherence to some of the less practical early requirements for computerized system validation such as requiring 100 percent code reviews for all computer programs. Unknowingly this was a good thing for both the FDA and industry as it highlighted practical issues with implementing the generalized guidances and allowed both field inspectors and industry to adopt specific positions in terms of demonstrating compliance. Also, both industry and the agency were beginning to integrate risk as a central part of the compliance argument.

Risk Management and Science

Risk has always been implicit in terms of the drug development compliance argument; the concept of risk has largely been applied to the quality decision-making process. Most quality professionals routinely use Acceptable Quality Levels (AQL) of establishing sampling and inspection plans, but cannot discuss the appropriateness of applying AQL to continuous rather than discrete variables. Similarly, introducing the concept of Lot Tolerance Percent Defective (LTPD) which focuses on consumer risk and is an alternative to AQL criteria, which focuses on producer risk, is often met with confusion and resistance. Largely the concept of risk has been associated with the quality disposition element of compliance and has not been broadly used as a tool for compartmentalizing and quantifying risk. That is not to say that risk management tools have not been used as part of the drug development or compliance process. Drug delivery companies, which routinely have a device and drug component to the product, have used Failure Mode and Effect's Analysis (FMEAs) as part of the device development process for years. Error allocation analysis has been essential to establishing release criteria between the device and drug sub-systems drug and the overall system. Many manufacturing companies have employed Statistical Process Control (SPC) on the shop floor. Clinical trials must be properly powered in order to support evaluation of the clinical outcome of interest. All of these tools help the organization to understand the limits of allowable variation and describe the potential pitfalls within the process. However, they have never been applied in a concerted manner across industry, and consequently represent the exception rather than the rule.

The FDA's rapidly diminishing ability to depend upon inspection and documentation as its backbone for risk management drove it to became more and more risk averse—making drug development costs skyrocket. The FDA argued that this was not the case, as the average time for NDA approvals was holding steady, partly due to additional resources the agency could apply to the review process as a result of Commissioner Kessler's Prescription Drug User Fee Act (PDUFA) fee paid by manufacturers upon regulatory filing. However, what was not being tracked were the additional hurdles being thrown at the industry at the IND level. APIs being applied in combination therapies or utilizing improved drug delivery platforms had been in the marketplace for decades, and were suddenly being asked to undergo additional carcinogenicity and mutagenicity studies in a desperate effort by the agency to drive their risk exposure as low as possible. The result was fewer and fewer new drugs coming on to the market with less time for manufacturers to recoup their investment. The other reality is this approach was not really driving down risk as it did not address the underlying sources of variation in these therapies which could represent a threat to the patient. The issues came to a head in 2004 when David Graham, an Associate Director of the FDA Office of Drug Safety, used the protection of the whistleblower act to address the public risk from Vioxx. Dr Graham had previously been successful in removing the unsafe drugs Omniflox, an antibiotic, Rezulin, a diabetes treatment, Fen-Phen and Redux, weight-loss drugs, and phenylpropanolamine, an over-the-counter decongestant, from the US market and in restricting the use of Trovan, an antibiotic, to use in hospitals. He also had a part in the removal of Lotronex, Baycol, Seldane, and Propulsid. Dr Graham cited the fact that the FDA succumbed to the industry pressures of a potential blockbuster drug and disregarded the warnings of its own scientist in terms of the safety risks.

The agency knew this could not continue and began a wholesale transformation for defining acceptable drug development. One based on scientific understanding and risk management as a foundation for demonstrating safety and efficacy. The FDA launched its Critical Path Initiative in an effort to "lean out" the regulatory approval process. Specifically the objective was to modernize the techniques and methods used to evaluate the safety, efficacy, and quality of medical products as they move from candidate selection and design to mass manufacture through better predictive and evaluative tools:

- innovative trial design, new statistical tools and analytic methods, use of modeling and simulation;

- establishing and qualifying predictive biomarkers for specific conditions;

- less stringent Current Good X Practice (cGxP)[1] regulations for IND exploratory studies and clinical trials.

In order to achieve this industry had to change the way it developed medical products and established Quality Assurance. To do this the agency looked at best

---

1   FDA compliance: x can mean Clinical, Laboratory, Manufacturing, Pharmaceutical.

practices across industries and around the world. It had participated heavily as part of the International Committee on Harmonization (ICH) of Technical Requirements for Registration of Pharmaceuticals for Human Use which had attempted to consolidate best practices across the industry. Out of this effort came several landmark guidance documents which became the foundation for most global regulatory philosophy. Three critical guidance documents were ICH Q8, ICH Q9, and ICH Q10. ICH Q8 provided guidance in terms of the Pharmaceutical Development section of an ICH regulatory submission. It indicated areas where the effective application of risk management and pharmaceutical and manufacturing sciences can create a basis for flexible regulatory. ICH Q9 provided guidance in terms of establishing a quality risk management framework. It provided guidance on the principles and some of the tools of quality risk management that can enable more effective and consistent risk-based decisions, both by regulators and industry, regarding the quality of drug substances and drug (medicinal) products across the product lifecycle. ICH Q10 attempted to integrate these two new concepts as part of pharmaceutical quality system.

With these best practices in place, the FDA issued a series of new guidance documents which were intended to integrate this new thinking. The first landmark guidance was issued in September 2003 by the agency and was entitled *Pharmaceutical cGMPs for the 21st Century—A Risk-Based Approach.* This guidance laid out the vision for the agency and attempted to describe the benefit to industry in terms of reduced regulatory burden and potentially faster time to market. This was quickly followed between September 2003 and March 2005 by a series of additional guidance documents:

- Guidance for Industry—Quality Systems Approach to Pharmaceutical Current Good Manufacturing Practice Regulations.

- Guidance for Industry Sterile Drug Products Produced by Aseptic Processing–cGMP.

- Guidance for Industry PAT—A Framework for Innovative Pharmaceutical Development, Manufacturing, and Quality Assurance: encourages industry to adopt online testing, characterization and control.

- Guidance for Industry Premarketing Risk Assessment: lays the groundwork for risk assessment process during product development, with emphasis on Phase 3 clinical trials.

- •   Guidance for Industry—Development and Use of Risk Minimization Action Plans: documents additional steps companies can take for patient risk that may not mitigated by traditional product labeling.

Together, these guidance documents represent a wholesale shift in quality thinking, one which is based on process understanding and predictability as opposed to product performance as the primary metrics for product safety and efficacy. Of these, the two primary considerations for the LSS professional are ICH Q8 and ICH Q9. ICH Q8 attempts to describe an approach to product development called QbD. In the QbD concept, the ability to demonstrate process predictability can be divided into three distinct phases. The first phase is the understanding the "knowledge space", which represents the total process space that could affect the process. From this exploration, the key variables which drive process performance are identified and the contribution to variability for each of these critical components is explored in order to establish the practical allowable variability of these variables as they affect process stability. This is called the "design space." Finally, limits are placed around these key variables which will ensure the process remains in control. This is called the "control space." Moving through this progression not only identifies the primary variables that are essential to process stability and product performance but have the added benefit of simplifying the downstream validation exercise. The concept behind QbD is illustrated in Figure 2.1.

Understanding this progression is essential to the LSS professional. Being able to summarize and portray the conclusions from an OpEx initiative must be scientifically defensible but also must be consistent with the final compliance argument.

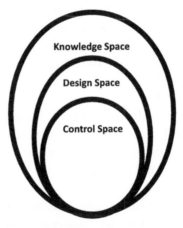

Figure 2.1   Process understanding through QbD

ICH Q9 fundamentally embraces the concept that every step of the drug development process involves some element of risk. A typical risk management framework is shown in Figure 2.2.

In order to effectively apply a risk management framework, the technical and quality organization must consistently apply risk management tools to support its decision-making process. Six Sigma and Lean Kaizen efforts are based upon a structured and measurable approach to process characterization and improvement. Some of the risk management tools that can be applied along with their strengths and weaknesses are discussed in the following section.

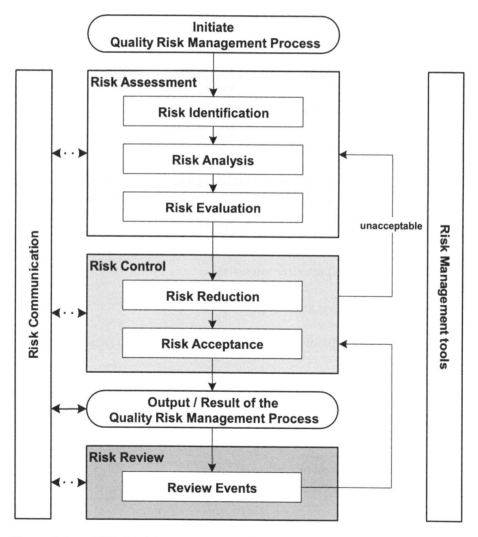

Figure 2.2   ICH Q9 risk management framework

Fault Tree Analysis (FTA)

FTA is a team-based method used to identify the causal chain that creates a hazard or a failure mode (effects are typically ignored). It represents the sequence and combination of possible events that may lead to a failure mode. Once the causes are identified, preventive action can be taken. It is composed of a series of Events and Gates. An "Event" is a cause or an effect. A "Gate" defines the conditional relationship between causes and effects, between x and y (what must happen for the effect to occur).

FTAs are often used:

- when conducting a risk analysis of a new facility or equipment;

- when multiple causes of a failure mode are suspected;

- when an interaction of causes is suspected;

- as an input to a FMEA;

- as an input to an experimental design to characterize a process or to determine variables and levels that will create a failure mode.

FTAs do have some limitations, they:

- are time and resource intensive;

- require expert knowledge of system under study;

- can lead to paralysis by analysis (infinite chains of cause and effect);

- require Microsoft Visio or other specialized software to document;

- are more useful as a problem solving rather than a problem prevention tool.

An example of a FTA is shown below in Figure 2.3.

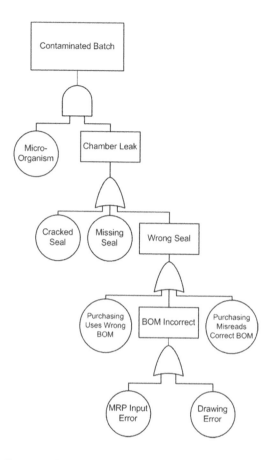

Figure 2.3   FTA analysis of a contaminated autoclave batch

Hazard Analysis and Critical Control Points (HACCP)

A HACCP is a method of identifying and controlling sources of variation in critical process steps that could lead to a hazardous condition. It is similar to a control plan. It cannot be used effectively without manual or automated process control methods, including SPC. HACCP has been recognized as a useful and effective risk management tool by the FDA for some time and became a mandatory component of the FDA's regulation for all seafood processing plants in 1997 and has been mandatory for canned food manufacturers since 1973. HACCP is intended to be a structured way to easily analyze and document risk within a process or piece of equipment. The basic steps for conducting a HACCP are:

- flow chart the process;

- identify any hazards that must be prevented, eliminated, or reduced to acceptable levels;

- identify variables that can cause the hazard (Design of Experiments or other tools);

- identify the critical control points in the process where those variables are impacted/affected;

- establish critical limits for those variables beyond which the hazard is created;

- determine and implement effective monitoring procedures at critical control points;

- determine corrective actions when monitoring that indicate that a critical control point is not under control (adjustment, maintenance, and so on);

- establish records of control and correction.

HACCP analyses are often used with new manufacturing processes or equipment.

The limitations surrounding HACCP analysis are:

- it requires excellent process knowledge to be effective;

- a FMEA should precede a HAACP to identify critical hazards/ failure modes (a HAACP could be an action to reduce the risk in a FMEA);

- it requires use of more complex statistical tools to be effective (Characterization DOEs, SPC, Capability Matrices).

An example of a typical HACCP form for a compression operation is shown in Table 2.1.

Table 2.1   Compression operation HACCP

Hazard	Critical Control Point	Variable to Control	Target Value	Acceptable Limits	Comments	Monitoring and Control Method	Corrective Action	Record
Tablet Breakage	Tablet Form	Comp. Force	15kp	13 – 18kp	Thickness and harness of tablet	Automatic tablet thickness gauge	Eject Tablet	On Line Batch Record

Hazard and Operability Method (HAZOP)

A HAZOP is a team-based risk management tool designed to identify hazards and deviations from normal operations, determine the hazard level, and brainstorm actions and recommendations to prevent the hazard from occurring, or to minimize its impact. In a HAZOP the key parameters of a particular process design are identified. In a pharmaceutical process this would be the Critical Process Parameters (CPP) that affect process predictability. For a chemical reaction vessel, for example, these parameters could be temperature, mixing time, pressure, and time. Then the HAZOP leader applies a series of guide words to the critical parameters and the team determines if any apply to the critical parameters in terms of a possible hazard. An example of a Guide Word table is shown below in Table 2.2.

Table 2.2   HAZOP Guide Word table

Key Words			
Key Word	Alternate	Definition	Example
No or none		The design intent does not occur	No flow
Less	Low	A quantitative decrease in the design intent occurs	Less pressure
More	High	A quantitative increase in the design intent occurs	More pressure
Reverse		The opposite of the design intent occurs	Reverse flow
Also		An unexpected activity, other than the design intent, occurs	Contamination
Other than		The activity occurs, but not in the way intended	Leak in pressure
Intermittent		The design intent is achieved only part of the time	Intermittent flow
Early		The design intent occurs before it is expected	Pressure before needed
Late		The design intent occurs after it is expected	Pressure after needed

The hazard severity and frequency are also assessed in order to establish the hazard levels. The risk matrix for this assessment may already exist for the process and product or may have to be developed by the team as part of the HAZOP exercise. An example of a risk matrix is shown in Figure 2.4.

Risk Level						
Frequency of Occurrence		**Severity**				
		(1)	(2)	(3)	(4)	(5)
		Catastrophic	Critical	Moderate	Marginal	Negligible
(A)	Frequent	1A	2A	3A	4A	5A
(B)	Probable	1B	2B	3B	4B	5B
(C)	Occasional	1C	2C	3C	4C	5C
(D)	Remote	1D	2D	3D	4D	5D
(E)	Improbable	1E	2E	3E	4E	5E

Figure 2.4   HAZOP risk matrix

An example of a HAZOP for a chemical reaction vessel is shown below in Figure 2.5.

HAZOP Template								
Process Function	Key Word	Deviation	Cause	Effects	Current Controls/ Safeguards	Hazard Level	Actions/ Recommendations	Record
Provide proper temp.	Low	High temp in blender	Thermo-couple failure	• Feed material #1 reaches decomposition temperature • Violent reaction with toxic gas generation • Personnel exposure/ injury • Equipment damage	Shutdown over temp sensor	1D	Replace with top quality thermocouple Maintenance verification of thermocouple accuracy weekly	Maint.

Figure 2.5   HAZOP analysis for a chemical reaction vessel

HAZOPs are often used for new or current manufacturing process or equipment and can be an excellent planning tool for maintenance groups.

HAZOPs have the following limitations compared to other risk management tools:

- risk level more subjective than FMEA;

- typically there is no reassessment of risk after risk controls applied.

Failure Modes and Effects Analysis (FMEA)

Probably the most prevalent risk management tool being used today by the pharmaceutical and biotech industry is FMEA. FMEAs can be used in many phases of the product development lifecycle. Typical applications of FMEA include System FMEA, Subsystem FMEA, Component FMEA, Equipment FMEA, Automation FMEA, Design FMEA, Process FMEA, Service FMEA, and Improvement FMEA.

FMEA is a team-based approach intended to ensure that sources of risk are identified and addressed through actions designed to:

- minimize the impact or severity of the risk;

- prevent the causes of risk from occurring; or to

- detect the risk early in its lifecycle to minimize its effect.

FMEAs are cross-functional and require representation from all departments that may be impacted by the risk. In most cases there is no representation from the ultimate stakeholder, the patient; however, the patient's impact is typically the key driver for the FMEA analysis. The FMEA is a living document and should be updated throughout the life of the product and the process development. Because an FMEA may determine that a facility, process, or machine design change is needed to reduce risk, it is best to conduct an FMEA as early as possible in the product development lifecycle, preferably during the product design and development phase.

While an FMEA is not really quantitative in its risk assessment capability, its strength comes from evaluating risk from a common platform. This allows multiple subject matter experts (SMEs) to participate and evaluate risk from a common perspective. Before beginning the FMEA the team must establish agreed-upon metrics, called ranking tables, for assessing risk severity, frequency of occurrence, and likelihood of detection. Examples of these tables are shown in Tables 2.3, 2.4. and 2.5.

Table 2.3   Severity ranking table

Effect	Criteria: Severity of Effect to Customer	Criteria: Severity of Effect to Manufacturing	Ranking
Hazardous-Without Warning	Financial loss > $1M Threat to patient safety	Inability to supply product Product recall Inventory hold (pause)	10
Very High	Out of compliance with regulations Financial loss between $.5M and $1M	Lose part of a batch Variability in supplier capability Inability to supply to the CMO Inability to maintain inventory	8
Moderate	Financial loss less than or equal to $.5M	Delays manufacturing by 1 to 2 weeks Affects manufacturing yield up to 10% Boosts standard costs by up to 10% Material scrap	6
Minor	Financial loss less than or equal to $100,000	Increased supplier diligence required Reprocessing / rework required Additional work (extra step) required Raw material defects Extra material required	3
None	No discernible effect	No discernible effect	1

Table 2.4   Occurrence ranking table

Probability of Failure Occurring	Possible Failure Rates	Rationale	Ranking
Very High: Persistent Failures	>10%		10
			9
High: Frequent Failures	5% – 10%		8
			7
Moderate: Occasional Failures	1%–5%		6
			5
	.1% – 1%		4
Low: Relatively Few Failures			3
	0% –.1%		2
Remote: Failure is Unlikely			1

Table 2.5   Detection ranking table

General Criteria	General Criteria Explanation	Inspect Types		Defective Product Criteria	Ranking
		Mistake Proofing	Manual Inspection		
Almost Impossible	Absolute certainty of non-detection		X	Cannot detect or is not checked	10
Very Remote	Controls have poor chance of detection		X	Control is achieved with indirect or random manual checks only	9
Remote	Controls have poor chance of detection		X	Control is achieved with on-line manual inspection only	8
Very Low	Controls have moderate chance of detection		X	Control is achieved with off-line manual inspection or charting methods, such as SPC	7
Low	Controls have moderate chance of detection		X	Control is achieved with manual on-line charting methods, such as SPC	6
Moderate	Controls detect and do not segregate or prevent defective product	X		Control will provide indicator (visual or auditory alarm) of failure after occurrence, but will not segregate defective product	5
Moderately High	Controls detect and segregate defective product, but do not prevent	X		Control will provide indicator (visual or auditory alarm) of failure after occurrence and segregate defective product	4
High	Controls detect imminent failure	X		Control will provide indicator of imminent failure with a visual or auditory alarm	3
Very High	Controls prevent and inform operator of corrective action	X		Control will detect cause, stop machine if immediate action not taken, and provide operator with information to adjust/maintain machine to prevent failure	2
Almost Certain	Automatic feedback control and correction	X		Control will detect cause and automatically adjust machine to prevent failure	1

Based upon these agreed-upon ranking tables the team will then step through each process and determine if there is a possible failure mode and assign a severity, occurrence, and detection ranking. These values when multiplied together comprise the Risk Priority Number (RPN). The RPN is used to prioritize mitigation actions for each identified risk. An example of an FMEA is shown in Figure 2.6.

Any OpEx initiative that does not start with a regulatory and compliance risk assessment has the potential to not only jeopardize the viability of the improvement but also could impugn product already in the field. These risk management tools will allow the organization to compartmentalize the risk so the data derived from the OpEx study can be used most effectively to achieve process stability and reduce compliance exposure.

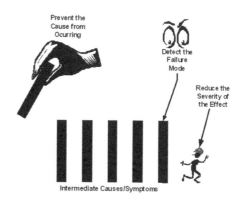

Figure 2.6   Example of FMEA

The Expanded Supply Chain: Managing the "Pharmerging" Markets

With the collapse of the global economy over the last 13 years the market has looked to expand the supply chain overseas in order to take advantage of lower cost capital and labor costs. The original countries of Brazil, Russia, India, and China, termed the BRIC nations, represented the fasting growing economies for the last five years. These countries were attractive to the pharmaceutical industry for several reasons. They each had a growing educated workforce which was essential to navigate the highly regulated drug development process, there was ready access to low-cost labor and capital and each one had governments that had identified the pharmaceutical market as a growth opportunity. Interest in expanding into these countries has been so significant by the pharmaceutical industry that they have been dubbed the "pharmerging" markets.

These markets represent a significant challenge for LSS initiatives because of the difficulty in establishing the true capability of suppliers and contract service providers overseas. Today, over 80 percent of the world APIs come from China and India. Almost 40 percent of the US markets' APIs come from these countries as well. With a standard cost at one-third that of US and European suppliers it is difficult to ignore the opportunity these markets represent. Several high-profile missteps have, however, brought home the risk management trade-off implicit in these overseas operations. The Heparin contamination with hypersulfated chondroitin sulfate in 2009 may have played a role in up to 131 deaths in the US alone, while exposing Baxter to billions of dollars of litigation. The FDA audit revealed a consistent pattern of fraud and lack of cGMP compliance. API suppliers are measured against the ICH Q7A guidance which defines best practice for development and quality systems. The transgressions for this supplier were so large the quality systems never existed. How a supplier could be qualified as part of a Quality Audit remains a mystery and underscores the need for controls beyond inspection and corporate policy. Understanding the supply chain thoroughly, from the manufacturer all the way back to the sub-suppliers, for the components of a drug is critical to understanding the sources of variation and assessing the short and long-term risk exposure from any improvement initiative. Often, the focus of risk lands on the technology components of a process leaving it vulnerable to the basic quality risks that underscore our assumption of process predictability.

Any Lean or Six Sigma initiative must consider the underlying variability at the supplier of not only the API but its excipients. Excipient manufacturers will state that the level of control demanded by many pharma companies exceed what can practically be guaranteed, however, the onus falls upon the drug manufacturer to ensure all critical tests and specifications are in place to ensure the raw material variability is in line with drug process requirements.

As with all improvement and stabilization initiatives there is never enough time or resources to do all the studies necessary to drive the risk of deviation or failure to zero. Using these risk management tools will assist the organization and the OpEx initiative in focusing on only those critical elements that are essential to providing project success, including in the definition of success the compliance and regulatory path forward.

Chapter 3

Lean in the Product Development Environment

Delay always breeds danger; and to protract a great design is often to ruin it.

Miguel de Cervantes

When people think of Lean the immediate response is to consider the benefits as they can be applied to the shop floor. The principles of Lean Manufacturing defined by Taichii Ohno, the creator of the Toyota Production System (TPS), have been successfully applied across industries. Despite the recent high-profile issues encountered by Toyota, the basic principles of Lean Manufacturing are still considered effective and relevant to all processes. The principles of Lean Manufacturing pertain to eliminating waste in a process. It does not matter if the process pertains to manufacturing or is transactional—the basic principles remain true. The revolutionary concept in identifying waste as defined by Lean is to focus on those activities that are valued by the customer, not the business. This is a revolutionary concept for many reasons. By focusing upon what the customer values, inefficiencies within a process which may have been developed as part of an evolutionary business practice can be scrutinized within the context of what the customer will value, not what the business thinks is important. As an industry, the pharmaceutical industry has not been very diligent in always focusing on what the customer values. This concept is called the Voice of the Customer (VOC). Lean uses the VOC to help define those activities within the process that are essential to delivering what the customer wants from the process. In the pharmaceutical industry this would be safe and efficacious drugs supplied at a fair price. This last component has been the most contentious issue for the industry. The impact to the US pharmaceutical industry from the passage by Congress of the landmark healthcare legislation is still unknown. However, it is safe to say there will be steady pressure on industry to keep the cost of new drug therapies down. The dilemma for industry is how to do this and not give away business performance in an increasingly competitive marketplace. Realizing the benefits of Lean requires changing the organization's paradigm for how processes are developed, implemented, and monitored. Part and parcel of this organizational transformation is the recognition that in order to move toward the goal of becoming a Lean enterprise it is essential to exploit all of the opportunities for increasing process efficiency in every facet of the business. This means leveraging the opportunities within R&D, Purchasing, Supply Chain, Quality, Regulatory, Engineering Finance, and Operations. An example of a Lean pharmaceutical organization is shown in Figure 3.1.

Figure 3.1 **The pharmaceutical Lean organization**

So when faced with a looming potential for increased pricing pressure, it makes sense to focus on the part of the business that is the largest component of the overall product cost. For the pharmaceutical industry the largest driver for pricing is the time and energy spent on product development. Even in the current weak economy the industry invests approximately 18 percent of its gross domestic sales revenue into R&D. Couple this with an average time to market of ten years or more, the notion of applying Lean principles within the development environment becomes compelling.

R&D versus Manufacturing

To be successful in applying the principles of Lean in the R&D environment it is important to understand the cultural and organizational differences between R&D and manufacturing. This is essential because the cultural drivers and metrics for excellence could not be more different between these two parts of the product development lifecycle. In manufacturing, any discussion surrounding efficiency and cost savings resonates not only with the leadership but also with the resources that are responsible for executing the manufacturing business plan. Greater efficiency should translate into greater effectiveness which means increased revenue and business performance. The R&D environment, however, is valued on its ability to invent, innovate, and deliver new and more effective drug therapies. The concepts of discovery and invention fall more easily into the creative bucket than the execution bucket. On average, a pharmaceutical company will identify and develop more than 300 different molecular entities in an effort to bring one product to market over a period of ten years. Few industries could tolerate such a deeply inefficient process. With such hugely disparate philosophies and cultures it is reasonable to ask if it is possible to apply the concepts of efficiency to the development process and reap the benefits of an integrated process at the commercial level. I believe it is possible but the approach to applying these principles must be communicated properly to be effective.

The Toyota Product Development System

One of the most successful applications of Lean principles within the development environment is again attributed to Toyota. The Toyota Product Development System (TPDS) is different to the Toyota Production System (TPS) and reflects the principles and values Toyota has used to build a reputation as the benchmark of automotive quality. Despite their recent high-profile troubles with the sudden acceleration issues in the Prius model cars, the principles of TPDS are proven. We will discuss some reasons why Toyota drifted from their recipe for success in our "Lessons Learned" discussion at the end of this chapter.

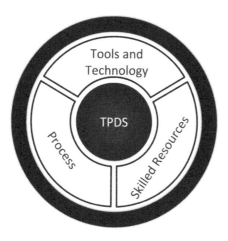

Figure 3.2   TPDS triumvirate

The TPDS is predicated on a holistic approach to product development which attempts to link the three essential elements of the development environment; the people, the development tools, and the process used for development.

Within this basic triumvirate there are 13 different elements which constitute the necessary framework for increasing product development effectiveness. These 13 elements are listed below:

TPDS—Process

1.   focus on customer-driven value and eliminate waste;

2.   front load the development process in order to explore and exploit the product design space;

3.   create a leveled product development process flow;

4.   implement standardized work practices where possible in order to maximize creative flexibility and establish predictable outcomes.

TPDS—Skilled resources

5.   implement a development head (chief engineer concept) that is responsible for integrating development from start to finish;

6.   organize the development effort to maximize functional expertise and cross-functional expertise;

7.   develop exceptional technical competence in all technical personnel (engineers);

8.   fully integrate suppliers into the product development lifecycle;

9.   embed systems to foster learning and continuous improvement;

10.   create a culture that will support excellence and relentless improvement.

TPDS—Tools and technology

11.   tailor your tools and technology to fit your people and processes;

12.   align your organization with simple visual communication;

13.   use powerful tools for standardization and organizational learning.

These principles reflect the doctrines of a consumer-driven industry in which the customer experience is a critical component of business success. This is not explicitly the case for the pharmaceutical industry where the customer driving many of the business decisions is shareholder value rather than the patient experience.

PRINCIPLE 1: FOCUS ON CUSTOMER-DRIVEN VALUE AND ELIMINATE WASTE

The challenge for the R&D environment is translating the Voice of the Customer (VOC) into terms which resonate in the discovery and development environment. Pharma in particular has reinforced the disconnect between the discovery and development environment and the commercial deployment phase of the drug development lifecycle. There are two major components to waste in the discovery phase of the drug product lifecycle. The first is pursuing molecular entities which do not hold promise in terms of their therapeutic indication. The second is the inefficiency of developing a product which has no commercial promise or does not justify the development cost and time based upon the market opportunity. A case in point, on average the pharmaceutical development environment will pursue 300 different molecular entities in parallel in the hopes of bringing a single new therapy to market. Few industries could tolerate such low success rates. On average the current time required for development and cost of new product development is ten years at a cost of 1.5 billion dollars. Given this astronomical baseline for product development the upside potential for forward thinking companies is enormous.

PRINCIPLE 2: FRONT LOAD THE DEVELOPMENT PROCESS IN ORDER TO EXPLORE AND EXPLOIT THE PRODUCT DESIGN SPACE

This refers to the old adage that 80 percent of your product development dollars are spent during the design process. While not true in the literal sense, the decisions made during the design phase of a product's development constrain the decisions that can be made downstream as the product moves toward commercial manufacturing. In the automotive world this means applying cross-functional expertise early in a development program to provide the maximum flexibility downstream and address potential design and manufacturing issues. This is easy to visualize when building a component for device. Similarly, in the drug development lifecycle the same principles apply. Early development focuses primarily on the functional capability of the molecule then drifts toward measurement through the pre-clinical process. As the product moves toward Phase 1 clinical testing, the intent is to evaluate whether there is any clinical efficacy in healthy volunteers. What does not typically happen at this stage is the integration of downstream requirements such as the robustness of key formulation components, the identification of key process variables, an evaluation of possible failure modes, or scalability and testing considerations. The irony is these considerations are often the impediments that drive up development cost and extend the development timeline, yet they could easily and quickly be addressed at small scale with little impact to timeline or development cost.

For example, a generic company may reverse engineer a product with two primary endpoints in mind: demonstrate bioequivalency to the brand product and be the first to file. Often this is a "hell bent for leather" undertaking with focused R&D groups bumping up against incredible timelines to file the ANDA. The result is often sub-optimized products which could have been engineered more easily and more robustly if a concerted cross-functional effort had been made to answer the required manufacturing and business performance questions early in the development lifecycle.

PRINCIPLE 3: CREATE A LEVELED PRODUCT DEVELOPMENT PROCESS FLOW

The key to creating a leveled product development process flow is knowledge management. While most companies do not want to constrain the creative process, successful development organizations learn quickly the key to flexibility in product design and development is structure, not the opposite. This may not seem intuitive, but Lean practitioners will tell you the purpose of a defined framework is for all the creative forces within the organization to focus on the problem and the objective of the process, not the process itself. Understanding, a priori, the metrics for success dictate the measurement of success. In terms of drug development this can be an evolving structure. Early in the discovery process, where molecular modeling and design may be the primary activity in the screening process, it is prudent to not over constrain the problem. Software packages designed to model molecules are only as good as the algorithms that built them, and most realize their practical accuracy is limited. However, as molecular candidates are identified, the establishment of more tangible success metrics is much easier. Defining a framework for measurement which harnesses the scientific horsepower of an organization will result in better and faster product identification. Establishing a cross-functional framework where experience and expertise are naturally tapped to provide input into the design and development process will ensure new products leverage lessons learned from past failures and successes. More importantly, it provides a measurement tool for management to gauge the effectiveness of its development process including the efficiency with which high-value decisions are made and the effectiveness of those decisions.

PRINCIPLE 4: IMPLEMENT STANDARDIZED WORK PRACTICES WHERE POSSIBLE IN ORDER TO MAXIMIZE CREATIVE FLEXIBILITY AND ESTABLISH PREDICTABLE OUTCOMES

This is an extension of the principles from the previous three principles. What Toyota strives to achieve is to promote a higher level of creativity by standardizing lower-level design tasks. The TPDS identifies three basic levels of standardization:

1. design standardization;

2. process standardization;

3. engineering skill set standardization.

"Design standardization" in the automotive world is achieved through modularity, common architecture, and where possible using shared components. The same principles can be applied in the drug development world. While there can be no doubt every development organization is looking for the next billion dollar blockbuster, it is preferable that this be achieved without spending a billion dollars in capital on new manufacturing testing capabilities. The truth is, most companies, by design or by accident, develop new therapeutic agents based upon a platform assumption. For example, it is unlikely a solid dosage company will develop an aseptic injectable as part of its primary development program unless it merits outsourcing the manufacturing. Immediate release dosage form platforms usually beget the same approach with the exception of enteric or controlled release technology that can leverage the existing manufacturing technology. So there is some standardization that takes place. The opportunity for development organizations is to standardize on the architecture for drug development adopting almost a drug delivery philosophy. Creating a portfolio of controlled release polymers, for example that work with a portfolio of molecular structures, allows each new product to build upon the lessons learned from the previous development program.

It also allows downstream support functions, like the materials management group, to negotiate better pricing based upon better volumes. The ability to established a stable platform with customizable performance capability leaves the development group with the best of both worlds, a stable platform that can tailor its performance to each new product requirements specification.

"Process standardization" refers to the ability to plug new products into an established characterized manufacturing framework. To a great extent this is what most development groups do but it is often done without process characterization in mind. The alignment between process development and product design must be done as part of the development process to be beneficial to the technology transfer process.

"Engineering skill set standardization" focuses on establishing a standardized pallet of product development skills across the organization to allow flexibility in developing high performance teams.

By developing this three-pronged approach Toyota is able to create a somewhat predictable development process in a marketplace that is constantly demanding a greater level of innovation and escalating levels of complexity. These principles will become even more critical to pharma as they embrace new, more complex delivery approaches provided by drug delivery platforms and nanotechnology.

PRINCIPLE 5: IMPLEMENT A DEVELOPMENT HEAD (CHIEF ENGINEER CONCEPT) THAT IS RESPONSIBLE FOR INTEGRATING DEVELOPMENT FROM START TO FINISH

This concept has been one of the most controversial elements of the TPDS approach. Fundamentally the concept is to create a formal focus within the organizational structure which is responsible for maximizing the collective learning of the organization. While simple in concept, as the complexity of product platform increases it becomes more difficult to stay on top of opportunities. Toyota's recent woes can be traced in some part to the failure of this concept to address issues before they become problems. Within the pharmaceutical development environment it is possible, however, to integrate the concept of maximizing collective learning as a formal component of the development lifecycle. This can be achieved through a development review committee. This is not a new concept within the R&D environment. However, what I am describing is a committee function which takes responsibility for the product development from R&D through commercial introduction at a technical level.

PRINCIPLE 6: ORGANIZE THE DEVELOPMENT EFFORT TO MAXIMIZE FUNCTIONAL EXPERTISE AND CROSS-FUNCTIONAL EXPERTISE

One of the largest challenges in implementing any kind of OpEx initiative is overcoming the silos which define the classical pharmaceutical organization. Approaches such as Lean Manufacturing and Six Sigma are predicated on committed and empowered cross-functional teams focusing upon projects and programs which have very clearly defined success metrics. That is not to say that pharma does not utilize cross-functional teams. Most product development programs use some fashion of product core team approach. However, the dynamic of these teams are very different from OpEx teams. Most development core teams have cross-functional membership in which each individual expertise has responsibility for delivering their component of the product development lifecycle. In OpEx teams the entire team has responsibility for delivering against the milestones and stage gates for the project.

As I stated before, one of the big challenges facing efficient and effective product development within pharma is that most R&D organizations are highly fragmented, making it difficult to identify the appropriate individuals to participate on project teams. This leaves open the potential for poor decision making from incomplete or incorrect information early in the development program.

One novel concept that has had some success in pharma R&D is the concept of *obeya* which literally means big room. The concept is an attempt to try and deal with shortcomings of the traditional product core team structure and pharma's organizational silos. The concept of a big room or "war room," is to have a location where the entire product cross-functional team meet daily. These are often stand-up meetings where critical problems are addressed as a team and decisions are made by the team in terms of how best to move forward. Within pharma, companies attempting to pursue a QbD approach are attempting to integrate downstream requirements into the design process. *Obeya* is the ideal of efficiently educating the entire product development team regarding the challenges faced by the organization at every step of the development process.

There are several benefits from the stand-up meetings. Besides gaining a better understanding of the challenges faced by each functional area throughout the product development lifecycle, the collaborative nature of the meetings helps to shorten the Plan–Do–Check–Act (PDCA) cycle time which exists even in the R&D environment. The hope is this will eventually lead to a quicker time to market. *Obeya* also attempts to break down walls between departments, and upper and lower management. Often times, upper management is not involved in the details of projects or efforts, leading to detachment and a potential breakdown in communication which can lead to delays in decision making. One final advantage of *obeya* is the location can move as the product moves closer to commercial introduction. This provides the opportunity for team members to get a direct understanding of the challenges and requirements for the product and process as it moves from invention to execution.

PRINCIPLE 7: DEVELOP EXCEPTIONAL TECHNICAL COMPETENCE IN ALL TECHNICAL PERSONNEL (ENGINEERS)

This may seem self-evident. Training has always been a big part of pharma's QMS. Most of us have experienced the panic of reviewing our progress against training goals for the year only to find you were short a few classes. However, the Lean concept of technical competence is much different. The concept here is to cultivate a culture of learning at the highest level. Prize expertise and capability and find a way to disseminate knowledge as broadly as possible. Within R&D there are many specialists but few have broad expertise. Principle 7 espouses a focus on creating a technical roadmap which will nurture your highest technical performers so the entire organization can bolster the tools in its toolkit.

PRINCIPLE 8: FULLY INTEGRATE SUPPLIERS INTO THE PRODUCT DEVELOPMENT LIFECYCLE

This is an evolving concept within pharma. We have typically viewed our suppliers early in the development process as having the capability to support whatever technical component or service we identified as critical to move the project forward. Lean's concept is to consider the supplier as you would an internal scientist or engineer. Recognize them for their expertise and attempt to build a level of commitment at the earliest level of the products development. Most organizations do partner with their suppliers as they move to the supply agreement and quality agreement phase of the product development lifecycle, but this is often not until Phase 2b or Phase 3 clinical evaluation. By now the decisions have been made and it is more difficult to seize opportunities that could dramatically reduce program risk downstream. For example, one manufacturer of highly successful class 3 medical drug/device combo hit a roadblock in their development program when the API supplier deemed the final product liability too large and would not sell them API to support their program. This organization, however, had built a strong relationship with several API distribution suppliers, even going so far as to involve them in the review and discussion of products in their development pipeline for manufacturing and capacity planning purposes. As a result, the development group was able to secure a long-term commitment from the distribution supplier and move their product to market.

PRINCIPLE 9: EMBED SYSTEMS TO FOSTER LEARNING AND CONTINUOUS IMPROVEMENT

Most large pharma have implemented some level of dashboard to track and report performance. Timelines within the development environment can span years and often become the last place the organization looks for information in terms of progress and continuous improvement opportunities. The foundation of OpEx is a culture of continuous learning and improvement. Organizations which have been successful in laying this foundation have extended the 13 principles of Lean to every facet of the development effort. Whether it is the catering that supports business meetings or the molecular discovery group that screens new potential molecular entities, there are metrics for performance and a feedback mechanism to foster improvement.

PRINCIPLE 10: CREATE A CULTURE THAT WILL SUPPORT EXCELLENCE AND RELENTLESS IMPROVEMENT

This is, of course, the ultimate goal. Toyota built its success by its unwavering belief in its core principles. Their most recent high-profile failures have also been because they strayed from those beliefs. To achieve this within pharma requires relentless leadership at all levels of the organization, from the CEO to the influence peddlers throughout the organization that are the voice of the organization's collective mentality and culture. Several firms have been able to achieve a measure of success through the application of focused programs such as "Right-the-First-Time," then embracing ICH Q8 and Q9 for its future development processes, but by and large this sensibility is the exception rather than the rule within the industry.

PRINCIPLE 11: TAILOR YOUR TOOLS AND TECHNOLOGY TO FIT YOUR PEOPLE AND PROCESSES

This has been the great weakness of our industry. We are enamored with the tools, techniques, and technology rather than the processes which underlie their application. Those that ever lived through the implementation of a new Enterprise Resource Planning (ERP) system can attest to the disruption and inefficiency borne by this new productivity tool. Applying technology accelerators to a fundamentally broken process does not enhance the process.

Lean focuses upon the inefficiencies with process steps. Within the development environment this means establishing a framework which will foster program predictability and ultimately effectiveness and efficiency. So understanding the current process is a critical first step to developing a better process. For example, if the analytical methods that are developed within the research environment do not seamlessly transfer to the development environment then a Laboratory Management Information System (LIMS) will be of limited value in terms of maintaining assay velocity through the lab. People and processes must always precede any implementation of tools or technology if the organization is going to reap any benefit from the implementation.

PRINCIPLE 12: ALIGN YOUR ORGANIZATION WITH SIMPLE VISUAL COMMUNICATION

This principle refers to the application of simple and universal tools that can be used to underscore the universality of the organization's business objectives, status, and issues. Some organizations have adopted a Balanced Scorecard approach or a Goals and Objectives (G&O) Cascade approach. With the development organization this can be a powerful tool, since, given the protracted timeframes of many of the development activities it is easy to lose sight of what you are doing poorly. Tying individual success to business success helps remind R&D of the consequences of doing their job poorly.

PRINCIPLE 13: USE POWERFUL TOOLS FOR STANDARDIZATION AND ORGANIZATIONAL LEARNING

One of the basic tenets of all OpEx philosophies is that through the broad application of standardization the end result is greater flexibility. These 13 principles are intended to emphasize that truth. By creating a framework, culture, and organizational structure that is geared to a prescribed roadmap, the risks of developing products which will perform poorly either in the clinic or on the shop floor will be greatly minimized.

There are some who believe implementing OpEx principles into the development process will squash innovation. Any organization that is not measured by effectiveness but rather timeliness would have the same worry. There have been many frameworks which have been successfully implemented in large pharma which have been able to demonstrate the benefits of a structured measurable environment. One large multi-national pharma company's R&D organization, for example, had grown exponentially over the years. Through acquisition of emerging technologies and participation with academia it had built a highly complex organizational structure which made direction and oversight nearly impossible at any strategic level. The results were multiple programs which were pursued in the early development program that had no clear end in mind. These little think tanks were successful in developing new therapies; however the commercial viability of many of these therapies was limited. Hundreds of millions of dollars went toward new therapies with a market size of tens of millions. This is not necessarily a bad thing if the business strategy is to pursue orphan drugs or specialty disease mitigation as part of some portfolio management strategy, but the development investment and market opportunity should not be a surprise.

Implementation Strategies

As always there is no magic bullet when it comes to achieving a paradigm shift in any organizational culture. The history of the organization, mores of the business environment, and leadership philosophy form the basis for how much effort it will take to change organizational thinking and how critical it will become to achieving business success.

In the development environment the metrics for success are different. For example, the costs-associated expenses are low compared to the investment required in resources and the lost market opportunities associated with long cycle times. Hence, if additional expense can relieve the bottleneck in a process this is a better use of capital. To do this the processes currently in place need to be well understood. The definition phase of the project is the most critical in terms of establishing a direction for changing organizational thinking.

While there is no single roadmap which can be applied to all organizations, Figure 3.3 illustrates an approach that has been successfully applied in some pharmaceutical R&D environments.

Let's examine each of these elements:

1. establish organizational charter;

2. develop current Value Stream Maps (VSMs);

3. develop intermediate and future state VSMs;

4. conduct organizational force field analysis;

5. develop a stage gate development process;

6. implement a knowledge management system.

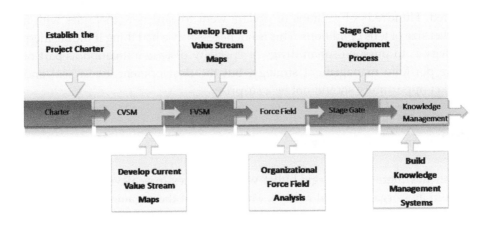

Figure 3.3 Lean R&D implementation roadmap

ESTABLISH ORGANIZATIONAL CHARTER

The chartering exercise does several things. First it defines the clear success metrics for the implementation. Most organizations contemplating applying Lean are interested in two primary success metrics. The first is reducing the time required to bring new therapies to market. The second is increasing the hit rate of new molecular entities in development. While both are essential to the long-term health and success of the business it is very difficult to pursue both as objectives of the transformation. My experience has been to focus on reducing the time to market as it is a more attainable goal which can resonate with an organization contemplating a paradigm shift. As with all Lean initiatives, the objective is a step function level change in terms of improvement. The goal must be tangible in order to support the investment and disruption that will follow from an organizational transformation. So, for example, a product development timeline where R&D may typically require four years to take a product through Phase 1 clinical trials may set a measureable improvement goal of two and a half to three years as an initial success metric. This goal can be extended further down the development timeline if development supports the product up through technology transfer and commercial scale-up.

DEVELOP CURRENT STATE VALUE STREAM MAPS (VSM)

When dealing with the eight Lean wastes the first step is to understand what you are currently doing. A value stream is a set of all actions, both value added and non-value added, required to bring product or service from inception to the customer. The first step in developing a value stream is to define who the customer is. Within the development environment the customer could be the patient, operations, clinical, regulatory, the Board of Directors, or the shareholders. Answering this question will define the boundaries of the high-level VSM. This exercise will start at the division level and drill down to the program level in order to establish a complete picture of the current process.

DEVELOP INTERMEDIATE AND FUTURE STATE VSMS

The process of developing a VSM identifies natural inefficiencies which are opportunities for eliminating waste. The intermediate and future state maps will come from iterative improvement against the optimization opportunities identified in the current VSM.

CONDUCT ORGANIZATIONAL FORCE FIELD ANALYSIS

The application of force field analysis as a tool for describing the dynamics of organizational change was first pioneered by Kurt Lewin. Lewin was a psychologist by training and he believed that when contemplating a change in organizational thinking the success of that change was a balance between the forces within the organization that wished to promote change and those that did not. Those wishing to promote change were considered "driving forces" for change while those against change were considered "restraining forces." To be successful, the driving forces must overcome the restraining forces if organization change was to be successful. The force field analysis (FFA) provides several key insights. It will identify the key issues involved in the balance of power. It will identify the key stakeholder that must be influenced and how they must be influenced and it will highlight the key opponents and allies of the proposed change. A simplified force field diagram is shown in Figure 3.4.

DEVELOP A STAGE GATE DEVELOPMENT PROCESS

The development of a stage gate program starts with intermediate VSM. This will set the framework for milestones which will represent the lost opportunity in TTM reduction. The elements within each stage gate will require broader

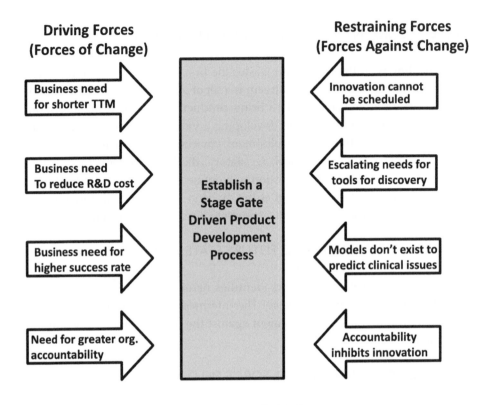

Figure 3.4   R&D organizational force field diagram

participation across the business spanning all departments across the development lifecycle through commercial introduction and possibly Phase 4 pharmacovigilence. Understanding the downstream requirements during the development process will allow the organization to leverage studies quickly and easily at small scale. The likelihood for resistance across the business will be less as addressing their concern early in the development process can only ease difficulties later on. However, it may require participation at an earlier level in the development process which may change the charter of some organizations. An FFA will be very helpful again in anticipating and addressing these challenges before reaching out across the business for these requirements.

IMPLEMENT A KNOWLEDGE MANAGEMENT SYSTEM

The final and most crucial step in reaping the benefits of Lean is the group efficiency in product development that will come from the gathering of holistic expertise. In order to truly push the development environment to the next level the R&D group must become efficient at failure. This may seem counterintuitive but true learning often comes from failure. What we want to do is disseminate that experience across the organization so we don't repeat the sins of the past. The same can be said for its successes. Some framework which allows development to utilize past failures and successes is essential to efficiently directing both the discovery process and development process. This should not be applied with the intent of stifling innovation, but rather focusing innovation, reducing the chance of failure with every product development iteration and harnessing the true scientific horsepower that often lies latent in the R&D environment.

A Word about Organizational Change

The science of organizational change is a large topic. The roadmap I have described is only the tip of the iceberg and to be truly effective should be combined with other tools and techniques to catalyze change. I will discuss these other tools as we move through the next few chapters with the intent of painting a picture of how Lean and other operational tools catalyze and support organizational change as it moves toward deployment.

Chapter 4

Applying Lean
in the Analytical Laboratory

An absolute can only be given in an intuition, while all the rest has to do with analysis.

Henri Bergson

The principles of Lean that have been so effective on the shop floor can have the same impact in the laboratory. As new therapeutic agents are identified, the complexity of characterizing and analyzing new molecular entities and biotherapeutic agents continues to escalate. Keeping pace with this escalating level of complexity are the myriad of analytical tests which are performed to ensure drug substance and drug product safety and efficacy. As organizations look further and further upstream for opportunities to reduce development risk, reduce time to market, and increase productivity, the role of the laboratory in the drug development lifecycle will vary. Understanding the role and impact of the laboratory is an essential precursor to optimizing its performance. Whether the laboratory is at the beginning of the product development lifecycle, supporting early molecule identification and formulation development, or at the end of the product lifecycle, supporting manufacturing and stability testing, the same considerations which impact a manufacturing process stream can impede the productivity of the laboratory. As with every Lean project, understanding the success metrics which are valued by the customer downstream is the first step in moving toward optimizing the process streams within the laboratory.

What is the purpose of the laboratory within the organization? What are the success metrics it is measured by and how are they measured? Understanding the demands placed upon and created within the laboratory are essential precursor steps to improving the effectiveness of the laboratory. I use effectiveness rather than efficiency since the laboratory is a service organization, and the focus of Lean is to maximize the value added activities within the value stream as defined by the customer of the value stream.

When applying Lean in the laboratory the thinking shifts from maximizing Process Velocity to maximizing Assay Velocity.

Laboratory Challenges

Laboratories, like any other process stream, contain both value added and non-value added activities. Depending upon the function of the laboratory in the product development lifecycle the distribution of responsibilities may change but the basic challenges are universal for all laboratories. For example, product development laboratories have to balance the need to support new product development and regulatory filing support with clinical supply release and stability while the commercial quality control must support new and commercial product release and stability testing. The reality is all laboratories suffer from the similar challenges to its effectiveness:

1.   **Highly variable lead times**
      - Many laboratories try group testing for equipment to minimize equipment setup, media, and sample preparation. This makes equipment capacity planning very difficult to do. This inability to forecast spills over into the resource planning function as well.

2.   **Highly variable incoming workload**
      - Most laboratories experience significant peaks and troughs of activity because of the highly variable nature of incoming samples. This is one area which benefits naturally from greater predictability on the shop floor and across the product supply chain. Understanding the drivers which affect this variability is central to establishing a more predictable laboratory operation. Without this the result is often missed delivery deadlines and extra resource requirements because of cascading inefficiencies in the execution of the laboratory.

3.   **Lack of resource planning system**
      - Many laboratories rely upon the laboratory manager or supervisor to assign responsibilities within the laboratory on a daily basis. Without a clear understanding of the effort and time required for each analytical testing value stream any resource decision is a guess at best.

4. **Poor laboratory layout**
    -   There is always some planning involved in the layout of the laboratory but typically the considerations are space and support utility requirements rather than optimal efficiency and minimum effort. This is compounded by disorganized laboratory supply schemes resulting in overstocking of unnecessary items and expediting critical items.

5. **Poor analyst cross-training**
    -   Most labs have a clear program for training and qualifying analysts on an analytical test. However, without a clear understanding of the testing demand and bottlenecks in the laboratory it becomes very difficult to effectively assign resources and prioritize testing.

6. **Too much work-in-process**
    -   There is no sample flow in the laboratory despite significant analyst effort to complete testing. This could be created by bottlenecks in data review, test approval, and out of specification (OOS) investigations as an example.

7. **Poorly implemented LIMS systems**
    -   The ability to realize the benefits of automated information systems is completely dependent upon how they are deployed in the laboratory. The lack of a well thought-out implementation plan will often result in an increase in the amount of work in the laboratory. Automation for the sake of automation will rarely realize the benefits it was intended for.

8. **Poor use of Lean tools**
    -   What makes Lean so universally acceptable is that the tools are simple, intuitive, and in most cases easy to implement. However, if these tools are not reinforced, measured, and maintained the realized and future gains in a laboratory will quickly disappear.

Lean Tools for the Laboratory

The framework for improving the efficiency of a laboratory is the same as can be applied on the shop floor. Figure 4.1 summarizes the components that are effective when applied in the laboratory setting.

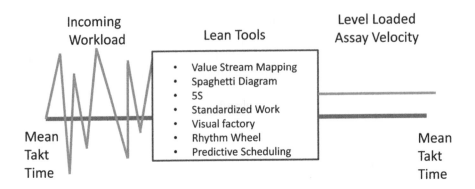

Figure 4.1   Applying Lean in the laboratory

The Linkage between Flow and Level Loading

The focus of any Lean exercise is to identify metrics and systems which will aid you in managing the laboratory to maintain predictable output based upon the business's needs. In manufacturing there is a key metric for stabilizing process streams concept called Process Velocity. Process Velocity is calculated as:

$$\text{Process Velocity} = \frac{\text{Total Throughput Time}}{\text{Value-Added Time}}$$

Process Velocity gives you an indication of how long it actually takes to complete the work that the customer is interested in. Ideally we would like the Process Velocity to be as low as possible.

In the laboratory the same cognate can be applied and is sometimes called Assay Velocity. The objective of the Lean exercise in the laboratory is to stabilize and optimize the Assay Velocity. To establish a stable and predictable Assay Velocity we need to flow the samples through the laboratory. You cannot flow samples through a laboratory unless the interval workload is level. One simple way to establish a flow is to use a tool like a Rhythm Wheel to stabilize the demand rate. A Rhythm Wheel repeats testing sequences in order to move the samples through all the required testing and data reviews quickly. This drives down the "throughput" time and incoming samples can then be held in a "leveling queue" at the start of the process.

A simple example of a Rhythm Wheel by Analyst is shown in Table 4.1.

Table 4.1   Rhythm Wheel by analyst

Test	Day 1	Day 2	Day 3	Day 4	Day 5	Day 6	Day 7	Total Weeks
HPLC	4	3	4	3	4	2	2	22
FTIR	1	1	1	1	1	1	1	7
GC				1		1	1	3
Dissolution	2	2	2	2	2	2	2	14
Analysts	7	6	7	7	7	6	6	

Samples in the leveling queue can be managed according to customer requirement using a system which targets when the test must begin. Rhythm Wheels work for simple processes. Where there are many sources of variation "Test Trains" can be created to break down the different test streams.

Product Development Quality Control Laboratory Case Study

Several years ago I was asked to lead a Lean exercise in a major biotech organization's Product Development Quality Control (PDQC) Laboratory. This laboratory resided in the R&D function of the organization and had been identified as the poorest performing laboratory in terms of on time delivery of analytical results. In addition the laboratory was working close to 24/7 in order to keep up with their workload resulting in overtime that was equivalent in cost to three full-time employees (FTEs). Lastly the demands of the workplace were taking a toll on the personnel within the laboratory and morale within the laboratory was very low. As with the Pilot Plant case study in Chapter 5, the project began with an overview of the principles of Lean Manufacturing, an overview of its tools and a discussion of the potential benefits from the exercise. The classes were interesting in that they included all of the laboratory personnel from director down to analyst as well as members of the organizations OpEx team that had asked to shadow our Lean project team. This training was customized to support the specific needs of the laboratory environment and set the stage for the group to move toward the chartering exercise.

Charter and Value Stream Maps Definition

Before beginning the chartering exercise the team spent time trying to better summarize what the responsibilities of the laboratory were from an organizational perspective. The PDQC laboratory had the following organizational responsibilities:

1.   product release testing;

2.   reference standards preparation;

3.   unplanned Phase 3 support;

4.   Phase 1 and 2 stability;

5.   toxicology.

The chartering exercise is designed to identify what the success metrics are for the project. This is most often tied to establishing what the customer values. For most operations, defining who the customer is and what they value can vary widely. When polled as to what was causing the greatest demands upon the lab, the team felt that product release testing was the biggest resource demand and should be focus of the charter. In an effort to confirm this perception the team reviewed the activity within the laboratory over the last year reviewing deliverables, timesheet summaries, and projects supported within the laboratory. To their surprise the data concluded that 93 percent of the demand in the laboratory was on activities two through five with only 7 percent allocated to product release testing.

For the first VSM exercise the scope was limited to Phase 1 and 2 stability support and unplanned Phase 3 support. The current demand for support was requiring significant overtime to maintain pace with the demand. If the capacity in the laboratory could be increased by 33 percent the overtime could be eliminated. The lifecycle for evaluation was defined as protocol generation through regulatory filing of the final report.

Value Stream Mapping

Value stream mapping is a Lean Manufacturing technique used to analyze and design the flow of materials and information required to bring a product or service to a customer. Value stream mapping has supporting methods that are often used in Lean environments to analyze and design flows at the system level (across multiple processes).

Although value stream mapping is often associated with manufacturing, it can be applied to any value chain. It is not required to use software to capture VSMs although there are several excellent solutions on the market that are cost effective. For this project eVSM was used to capture the current and future state maps for the laboratory. For the VSM exercise the team started with the input to the process and described the time and steps utilized in executing the Phase 1 and 2 stability program and the unplanned Phase 3 analytical testing support. Team members were given four types of symbols to use and the VSM was developed manually. The symbols used by the team are shown in Figure 4.2.

Process Step   Issue   Brainstorming   Process Input

Figure 4.2   Symbols used for VSM

The data were further refined into four categories of activity:

- process input;

- value added activity (VA);

- business value added activity (BVA);

- non-value added activity (NVA).

For every process step the cycle time was captured. If there was a changeover time associated with the step this was also captured within the process step. Processing time between steps was also noted and captured on a timeline at the bottom of the diagram. Most VSM analyses evaluate process steps by value added and non-value added activities but do not consider BVA. This is a key consideration when dealing with regulated processes such as GMP processes. Business value added process steps reflect activities which are either required for compliance or are endemic to the process. For example the "checked by" signature on a Manufacturing Batch Record (MBR) or laboratory summary sheet may not be value added from the customer's perspective but is required to demonstrate GMP compliance. Therefore this is not something the team should consider eliminating unless it is to replace it with a surrogate solution such as electronic signatures. The data were captured in eVSM along with all key process attributes for each step. The eVSM summary for the Phase 1 and Phase 2 stability VSM is shown below in Figure 4.3. Based upon the current state value stream map (cVSM) it is possible to capture the current state operations metrics.

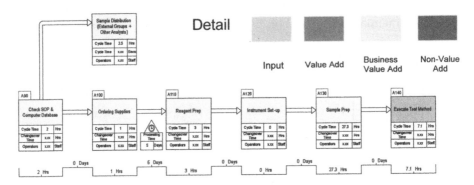

Figure 4.3   The eVSM summary for the Phase 1 and Phase 2 stability VSM

Current State Metrics

There is an old saying that describes the foundation of any improvement initiative: "If you can't measure you can't improve it." In order to achieve a 33 percent reduction in overtime it was essential to find a 33 percent improvement in capacity in the laboratory. The first task was to determine what the demand rate was in the laboratory. On the shop floor, this would be the rate at which material was required to meet the market demand. In Lean we call this the Takt time. The Takt time is the cycle time that sets the pace for the manufacturing process. In the laboratory this translates into the Assay Velocity through the laboratory. The Assay Velocity is defined as the rate at which an assay must be completed per assay condition. Takt time can be calculated as follows:

$$T = \frac{T_a}{T_d}$$

Where:

T = Takt time, (for example, minutes of work/unit produced)

T_a = Net time available to work, (for example, minutes of work/day)

T_d = Time demand (customer demand), (for example, units required/day)

For the PDQC laboratory the Takt time was determined to be 0.79 days/condition.

After completing the cVSM exercise the baseline metrics for the current operation were established and are shown in Table 4.2.

Table 4.2   **Baseline metrics for the current operation**

Metric	Current State Phase 1&2 Stability
Takt Time	0.79 days/condition
Cycle Time (CT)	190 hours
Lead Time (LT)	183 days
CT/LT Ratio	13%
Value Added Ratio	1.3%

So based upon the current state, the process tasks require 190 hours to complete, however because of inefficiencies in the overall process the actual time required to complete the tasks is 183 days. This translates in only 1.3 percent of the time spent during the process as being value added from the customer's perspective.

Laboratory Operation—Focus on Improvement

Lean professionals will recognize that a value added ratio of 1.3 percent is not unusual for large pharma and biotech. However, the laboratory team was shocked. We were nowhere near our target Takt time of 0.79 days/condition. It was important that the team recognize that this metric did not imply that they were not working hard, only that the current processes in place do not allow them be efficient and effective. With this premise the next step was to try and understand what could be done to move toward a more efficient laboratory process. To do this it was important to understand how the laboratory operated currently before a future state map could be developed.

The current laboratory was cramped and divided into three areas separated by corridors and doorways. There were nine analysts which ranged from entry-level analysts with three years' experience to senior analysts with over 15 years' experience. Two supervisors were in charge of the laboratory and one director oversaw the entire laboratory function. The laboratory did have a LIMS but the equipment was not on a network and the data were not automatically captured by the system. Every analyst had a notebook which they used for each assay performed which required pasting data into the book prior to the review. All analysts were expected to be able to run every assay after they had been qualified on the equipment and method. Because they supported regulatory filings and NDA/BLA submissions, all data had to be manually entered into the LIMS in order to generate the reports necessary. At the time of the project the supervisors were approximately nine to ten weeks behind in approving notebooks. In terms of stability pulls some sample pulls were four to six weeks behind the stability program schedule.

Workflow and Space Utilization

With the cVSM complete the team did a laboratory walk through to review equipment and material storage. The laboratory had several large double door incubators which were used for stability studies. These were filled to the brim with stability samples. Each sample was labeled but beyond that there was no attempt made to keep similar stability samples together. In addition there were a large number of stability samples from early development programs over the years which were still in the incubator. This was because it was unclear who was responsible for discontinuing the studies. Complicating the laboratory flow was the current practice of receiving stability samples on a weekly basis in the entry of the laboratory. Currently, once a week, two to four large insulated shipper containers would be received with stability samples which needed to be labeled and placed into the appropriate incubators within 24 hours of receipt. When this happened, all passage was blocked for a day to one-third of the laboratory as samples were removed, staged, and labeled.

Currently there were limited laboratory operation Standard Operating Procedures (SOPs) in place. Most of what was performed was based on tribal knowledge. The supervisors became the gatekeeper for all decisions regarding laboratory operation.

Managing the Workspace and Inventories

During the walkthrough the team spent time walking though the tasks for each assay and laboratory function. Most analytical tests required some element of sample preparation or upstream testing and verification. Each analyst was asked to develop a spaghetti diagram for each analytical test and for each laboratory function, such as labeling incoming stability samples. Creating a spaghetti diagram is the visual creation of actual flow. The keyword here is the "actual flow," not what it should be or is perceived to be. It is a snapshot in time so it may not include all what-if and special scenarios but is representative of the normal operation of the process. Spaghetti diagrams can be created for product flow, people flow, and data flow.

Typically, the recorder of the spaghetti diagram would also use a rolling measurement tool to capture the actual distance covered by the analyst during the execution of the process step. This would provide one metric which could be used to measure the improvement in the process operation after the Lean exercise. In this case, the team wanted to get a glimpse of the process primarily and was not concerned about the improvement metric. Rather the focus was on achieving the 33 percent reduction in resource utilization. As can be seen the analyst was forced to cover a great deal of the laboratory in order to execute the analytical test. Spaghetti diagrams for the other analytical test stations revealed similar inefficiencies.

5S and the Visual Factory

Since the laboratory was an existing space, the operation had to be built around the current configuration for the rooms. With this comes the classic pitfall of any evolving operation, that is, hidden inventory and non-optimized process workstations. With the spaghetti diagrams in place the team conducted a large 5S exercise. 5S is a Lean workplace organization methodology based upon the philosophy of ensuring the workplace is neat, orderly, and efficient with the necessary tools available to efficiently execute each process task. The Japanese terminology for 5S is *seiri, seiton, seiso, seiketsu,* and *shitsuke* which translated means sort, set-in-order, shine, standardize and sustain. Each workstation was evaluated against a predefined set of standards and assigned a value against each of these elements. The list of evaluation criteria are shown in Table 4.3. Further each workstation was marked for all components necessary to complete the assay.

Table 4.3   5S evaluation criteria

5 S Rating Chart

Date: _____   Department: _____

Shift: _____   Rating By: _____

	5 Points World Class Environment	4 Points Excellent	3 Points Good	2 Points Marginal	1 Point Poor	0 Points Unacceptable
SORT	Every item in the area is needed for the work of the area. Consider all items: materials, tools, equipment, furniture, documents.	Minimal clutter, but a small amount of unneeded material, tools, documents or general items. Parts & Work in Progress (WIP) are at standard levels.	Either the bench, shelves or cabinets at the station have some unneeded items, such as tools, documents or general items. Excessive parts & WIP without the department.	Some clutter & unnecessary items. There is no unneeded furniture or equipment, benches, shelves & storage cabinets have unneeded and/or excess items.	Generally cluttered. Quite a few unnecessary items on the bench, shelves or cabinet. Some unneeded furniture, equipment & general items. All emergency equipment accessible.	Station is cluttered & unsafe. Many unnecessary items on benches, shelves, cabinets & floor. Much unneeded furniture & equipment. Emergency equipment blocked.
SET-IN-ORDER	Station very well organized. There is a designated or labeled location for all items. Locations are convenient & items are stored properly. Equipment & aisles have designated markings. Excellent visual controls are in place. Every production part is labeled with part no.	Station well organized. Most items have a designated or labeled location. Although not perfect, most locations are good & items are stored properly. Equipment & aisles have designated markings. Many visual controls are in place.	Station well organized with the exception of a few problems. Most items have a designated appropriate location & most tools are stored properly. All aisles are marked, but some equipment is not. Many visual controls are in place.	Station moderately organized. Many key items have a designated location, but some do not & some items are improperly stored. Equipment & aisles nay not be clearly marked. There may be some visual controls in place, but no measurement board really visible.	Inadequate organization. Many key items have no designated location. Shelves & drawers are properly labeled. Aisles or equipment marking are not clear or straight. Very few visual controls or instruction are evident. Safety equipment is not designated. No measurement	Poor organization. Little indication for a designated place for anything. Items improperly stored on the floor & benches. Trip hazards. Furniture & tools located far from where needed. No current visual controls, signs or labels.
SHINE	Every item in the station is in "like new" condition. Oil/water socks are not required. Floors shine & baskets are emptied daily.	Every item in the station is clean and functional. Floors are clean & wastebaskets are emptied daily.	Most items in the area are clean & functional. No debris in part containers. Floors are pretty clean & baskets are emptied.	There is some clutter. Some items are not clean and/or swept. Some oil/air/water leakage in pan. Floors or waste containers are marginal.	Area needs to be picked up. Items have not been cleaned in some time. Significant oil/air/water leakage. Floor is dirty and/or waste containers are full.	Area is very dirty. Furniture, equipment, shelves, fixture tools are coated with oil and dirt. Possible health & safety hazard exist.
STANDARDIZE	There is a 5S schedule/checklist posted & is followed daily. Machines have a PM & Predictive Maintenance process being performed. All metrics/charts in the area are current with improving trends.	There is a 5S schedule/checklist posted & it seems to be followed. All required PMs have been performed & log is up to date. All metrics/charts in the area are current & show improving trends.	There is a 5S schedule/checklist posted & it seems to be followed. All required PMs have been performed & log is up to date. All metrics/charts in the area are current.	Although a 5S schedule/checklist is not posted, it is available. All required PMs have been performed & log is not up to date. Metrics/charts in the area are current.	Although a 5S schedule/checklist exists for the station, it is not posted. All required PMs have been performed & log is not up to date. Metrics/charts in the area are not current.	There is no evidence that a daily 5S schedule/checklist exists for the station. All required PMs have not been performed and/or log is not up to date. Metrics/charts in the area are not current.
SUSTAIN	Clearly, all levels of the organization are dedicated to sustaining the 5S.	There is strong interest throughout the organization to sustain 5S.	Management shows commitment to sustaining 5S.	Shop supervision shows commitment to sustaining 5S.	Some commitment in the organization to sustaining 5S.	There is no commitment within the organization to sustaining 5S.
			Score _____			of 25 points

Basic scores before the 5S exercise averaged three out of 25. Although it was not possible to reconfigure the laboratory by moving walls it was possible to try and address the issue of hidden inventory. All laboratories have cabinets and drawers. The team emptied all drawers and consolidated material. The laboratory uncovered dozens and dozens of boxes of laboratory materials for products that were no longer made or programs that failed. To reinforce the fact that hidden inventory was not desirable all of the cabinet doors were removed from each laboratory bench. This was not as bad as it sounds. Most of the analytical equipment required computers which were taking up valuable laboratory space. With the doors removed the team had holes drilled into the laboratory benches and placed the computers underneath the laboratory bench. All free-standing monitors were mounted on articulating arms so they could be lifted out of the way when not in use. The team then rearranged the laboratory based upon the spaghetti diagrams to minimize the travel, as best as they could in order to execute each task.

In addition to obsolete laboratory supplies the current laboratory supplies, including chemicals, were stashed all over the lab, many of which had expired and had never been opened. The team adopted a First-In-First-Out (FIFO) philosophy for all chemicals. A supermarket was created at each workstation for the chemicals necessary for each analytical test or operation, with the oldest material being placed first. All remaining chemicals were placed in cabinets in alphabetical order. Under each container was placed a green dot or a red dot. The oldest material was placed in front for each chemical and had a green dot under it. Based upon the usage a series of green dots were followed by red dots under each container. Once the red dots appeared it was time to reorder chemicals. The laboratory had a central service individual that picked up all inventory orders for all labs. With all inventory now visible, it was possible for this individual to come by once a week and instantly determine if new chemicals were required. Finally all areas were labeled above to identify their function and on each drawer to identify their contents.

After an initial 5S exercise the scores rose to 13 out of 25. In order to reinforce the necessity of maintaining an orderly workspace each of the analysts were assigned an area of responsibility. The baseline score of each area was posted along with the analyst responsible for it. By the end of the second week of implementation the average score had risen to 19 out of 25. The available space and visual appearance was noticeably better and the analysts agreed it was a much better environment to work in.

The cultural transition was the largest paradigm shift to overcome. All analysts like to have their own calculators, pipettes, and HPC columns. It has always been part of the laboratory sub-culture and has been associated with an analyst's ability to have faith in the equipment they are using since they are the only ones that use it. The reality is it creates excess inventory and inefficiency. At the end of the 5S exercise each analyst was given their own calculator as a reward for weathering the storm.

Data Bottlenecks

The cVSM identified a number of bottlenecks in the laboratory. First the two supervisors were not able to keep pace with necessary data reviews in a timely fashion, resulting in major delays to approved data and the laboratory falling behind in terms of the stability program. In response the laboratory created an approved list of data reviewers by analytical test. This allowed other senior analysts to do the review and approve the notebooks. The second bottleneck was the LIMS system. Since the equipment was not tied to the unit the analysts had to do double duty entering the data into the system and recording the data in notebooks. The team challenged themselves to see what the value add of the paper reporting system was versus the electronic database system. In the end, the customer doesn't care how they get their data but they do need accurate data that is delivered on time. The laboratory made the decision to develop a migration plan where all final data would be in the LIMS system and eliminate the paper reports. Once the team committed to the transition it was possible to work through the backlog and establish metrics for data entry going forward.

Scheduling

The laboratory was moving in the direction of standardized work practices but there was no system in place for scheduling activity in the laboratory. Because of this, it was impossible for the supervisors to know where they were in terms of their commitments or for analysts to know where other analysts were in their commitments. Consequently, last minute requests were tremendously disruptive to the labs productivity and Takt time. The laboratory supervisors and senior analysts were challenged to come up with a solution for scheduling activity in the laboratory and providing feedback on progress.

Previously the Takt time had been established for the laboratory at 0.79 days/operation. This Takt time represents the target time when the activity in the laboratory is level loaded.

As we discussed earlier, one approach that can be effective in determining a schedule is a Rhythm Wheel. A Rhythm Wheel is a fixed repeating sequence of tests that is designed to meet "leveled demand," and achieve consistent, repeatable "through-put" times while balancing daily workloads. The sequence repeats at a fixed internal which, for individual Rhythm Wheels, can vary from hours to weeks. Individual test runs are placed in the sequence in order to balance daily workloads and avoid equipment conflicts and so on. An example of a Rhythm Wheel for the laboratory is shown below in Table 4.4.

Table 4.4   Rhythm Wheel for PDQC laboratory

Assay	M	T	W	Th	F	S	Su	Total
HPLC	5	6	5	6	5	1	0	28
UV		1						1
pH	1	1	1	1	1	1	1	7
TLC	1							1
FTIR	1	1	1	1	1	1	1	7
PCR	1			2				2
Analysts	9	9	7	10	7	3	2	47

This fixed pattern repeats each week and can deal with the average expected sample every week.

The Rhythm Wheel gave a framework for the demand in the laboratory, but it did not provide any feedback to the supervisor as to the status of each analyst's workload in the laboratory. Since the laboratory supported stability programs in addition to other activities, we needed a system which would allow the supervisor to intelligently choose the right analyst for unplanned activities. If, for example, a typical stability time point allowed 30 days to complete all of the analytical tests per condition, it would be better to choose an analyst at the beginning of the 30-day period rather than at the end so as to not miss the time point for the program. The laboratory settled on a simple color-coded system of folders for all products which needed to be tested that week in the laboratory. An example of the visual filing system is show below in Figure 4.4.

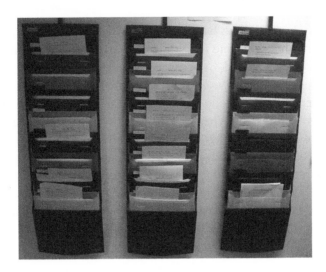

Figure 4.4   **Lab scheduling visual status board**

As each assignment is completed the folders are turned over so it is apparent what is in process and what is complete. After the first month of implementing the new system the time to completion was reduced from 30 days to 21 days.

Future State

With the data gained from the 5S exercise and the scheduling exercise the team had a very good understanding of the challenges in the laboratory and reconvened to try and map out a future state map. The team was allowed to include policies which were outside the view of the laboratory but impacted the laboratory from upstream and downstream organizations. Specifically, the greatest opportunity for improvement was in the product development group. The product development group was upstream of the PDQC group and was responsible for formulation and method development. The challenge facing the group was that the methods being developed more often than not did not transfer to the commercial equipment being used by the PDQC lab, leaving the laboratory to debug and troubleshoot the methods. If some accountability could be increased to the upstream group the PDQC group could stabilize its resource loading. Based upon the improvements in the laboratory and the future Value Stream Maps (fVSM) exercise the laboratory was able to make great strides toward the objective of increasing the laboratory capacity by 33 percent. The resulting metrics at the end of the Lean exercise are summarized in Table 4.5.

Table 4.5   Resulting metrics at the end of the Lean exercise

Metric	Current State Phase I and 2 Stability	Future State Phase I and 2 Stability
Takt Time	O.79 days/condition	O.79 days/condition
Cycle Time (CT)	190 hours	158.5 hours
Lead Time (LT)	183 days	67 days
CT/LT Ratio	13%	29.6%
Value Added Ratio	1.3%	12.5%
Percent of Goal	–	71.5%

Conclusion

The Lean Kaizen had increased the value added ratio from 1.3 to 12.5 percent and reduced the lead time from 183 to 67 days. Although the target improvement had not been fully attained at the end of the exercise the laboratory was well on its way. The additional recommended organizational improvements outside the laboratory had the potential for dramatically impacting the laboratory's productivity. As the laboratory refines its scheduling the productivity and predictability of the laboratory resource utilization will only go up. Most importantly, the laboratory now had metrics for measuring laboratory performance it could use to address future analytical demands on the operation. Other than the eVSM software, the Lean exercise did not require any special software or complex analytics. By sticking to a clearly defined business imperative and success metric the laboratory was able to significantly alleviate the overtime demand in the laboratory and improve the productivity of the laboratory while improving the quality of life of the analysts.

Chapter 5

Applying Lean in the Pilot Plant

Every well-built house started in the form of a definite purpose plus a
definite plan in the nature of a set of blueprints.

Napoleon Hill

Why worry about the efficiency in your Pilot Plant? While the basic tenets of Lean may have found their genesis in understanding and controlling the efficiency and waste on the shop floor, the impact of inefficiency upstream of the commercial shop floor can have just as large, if not larger impact on your organization's business performance. As we discussed earlier, when considering the application of Lean in the product development environment, any activity that can slow time to market or add risk to the potential success of a new molecular entity represents a significant threat to business performance.

The activities which take place at small and intermediate levels are taking on a new level of importance with the issuance of the FDA's new guidance on process validation in 2011. The guidance emphasizes the importance of process understanding which typically takes place during the development stage, calling this stage of process learning "Stage 1 of Process Validation." So the ability to effectively and efficiently utilize a Pilot Plant becomes more and more important as products move through the development pipeline. The role of the Pilot Plant in the drug development process is especially significant for biotech products. Unlike small molecule drug products, the complexities of the control strategy for a biotech process often rely on intermediate level processing and modeling. In Chapter 1 we discussed the basic steps in moving a molecule from discovery through FDA approval and commercial introduction. For biotech products the role of the Pilot Plant has historically been a central piece to managing the risk from premature capital investment and uncertainty as the product moved from Phase 2 to Phase 3. Today the rapid adoption of single-use disposable technology has made the risk profile somewhat more manageable but there still is no suitable substitute today for large-scale fermentation and microbial processing other than hard-piped facilities.

The challenge when applying Lean in the Pilot Plant is that the thinking must shift from product specific, process stream-based optimization to optimization of the key processes around each unit operation. Understanding what process stream to optimize is often the key success factor in bolstering productivity in a Pilot Plant setting.

Framing the Problem

What is the purpose of a Pilot Plant? Some may say it is a flexible platform for prototyping new processes and formulations. Others may say it is a center for building clinical supplies. Others may say it is the linchpin in the overall product development supply chain, feeding ongoing clinical trials while supporting pure research in the pursuit of identifying the next blockbuster drug. Depending on who you talk to, the role of the Pilot Plant can run the gamut from trivial to an essential stepping stone to success within the drug development process. With such a disparate perception of the role of work center in the overall drug development lifecycle it becomes essential to understand what the inputs and outputs are to an organization that must be studied and optimized. As with Six Sigma, it is exceedingly difficult to optimize multiple key objectives. Choose a primary output and potentially a collateral output variable to measure is essential if the optimization effort is to be successful. Otherwise it becomes very difficult to make the necessary trade-off analysis that is essential to optimizing the main process steam. To illustrate the application of LSS principles in this setting I will discuss a biotech Pilot Plant case study, however the approach and considerations would apply equally to a small molecule Pilot Plant.

Cell-Fermentation Pilot Plant Case Study

Several years ago I was asked to lead a Lean exercise in a major biotech organization's Cell Fermentation Pilot Plant. The business issue that needed to be addressed was a shortfall in capacity. The Pilot Plant was already running seven days a week and based upon the projected product development plan the Pilot Plant was projected to fall at least 20 percent short of the necessary capacity to accommodate new products in development. The Lean assignment included training as well as deployment. This was to be run as a Lean Kaizen with an estimated time commitment of the Kaizen team of no more than five days to identify and evaluate the improvements.

The project started with a high-level overview of the basic principles of LSS. The classes were a diverse mix which included key contributors to the operations and spanned the entire organizational structure, from vice-presidents to operators. This was significant because, without a rudimentary understanding of the philosophy and methodology of Lean, it would be very difficult to garner the necessary resources, expertise, and decision-making hierarchy necessary to effect change. It was also evidence of the top-down commitment from leadership to the success of this project. This training was customized to support the specific needs of the Pilot Plant organization and set the stage for the capacity improvement work.

Charter and Value Stream Map Definition

As with all Lean projects, we started with the chartering exercise to define the scope and success metrics for the project. The Pilot Plant Services (PPS) organization consists of two unit operations, cell culture and purification. The cell culture group managed both small-scale (2–10L batch size) and large-scale (100–1000L batch size) process streams. The large-scale operation was capable of supporting both bacterial and mammalian cell process studies in the same facility. The PPS group's responsibility was to execute and gather data per the predefined protocols developed by the process development group. To facilitate the chartering discussion the team had been presented with the following basic questions in advance of the meeting to help establish a charter:

1. What is the function of the Pilot Plant in terms of the drug development process?

2. What do the PPS group's customers value?

3. What is the problem we are trying to solve?

 Three seemingly innocent questions spawned nearly a week of discussion by the group, its customers, and management. The culmination of the discussion was the following. The PPS group is a service group that is responsible for the execution of development, scale-up, and process characterization studies. They do not participate in the design or analysis of the data, simply the gathering and transmittance of data. They are solely responsible for the operation of the planning, scheduling, and execution of studies within the Pilot Plant. After much debate it was agreed that, despite all of these functional activities within the business unit, the key product which was valued by the organization was data. Ironically, in looking at all of the systems that had been created within the group, the gathering and transmittance of data had little or no system's definition around it.

The capacity optimization work began with basic training in developing a VSM of the selected operation. This value stream mapping effort focused on the establishment of a cVSM as an assessment tool for identifying Value Added (VA), Business Value Added (BVA), and Non-Value Added (NVA) activities in the process. A Lean Kaizen team was established with the two supervisors from the organization acting as the team leads. The team agreed to work on the small-scale process stream since it constituted the largest capacity utilization for the plant and impacted the greatest number of internal customers.

The metric for project success was operational "capacity" measured as the number of small-scale experiments that could be run in a fixed period. The targeted improvement was set at a $\geq$ 20 percent increase over the established capacity. The VSM effort began with the team touring the facility to ensure all participants had a common understanding of the process. The tour identified several opportunities for applying some basic Lean tools including 5S, Visual Factory, Mistake Proofing, and Rapid Changeover for driving efficiency improvements.

Rudimentary baseline measures were available for the plant operations, including batch yield, batch failure rate, failure root cause analysis, and lot throughput on a monthly basis. The plant averaged four to seven studies a month, with each study typically consisting of between 11–19 experimental runs. The largest contributing factor to lot failure were the small-scale fermentors, called *Miniferms*, due to contamination, which averaged 9 percent on an annual basis. Although the operation tracked performance data there was no detail regarding cycle times for each process step, making it impossible to develop the cVSM.

In the absence of detailed unit operation data, the team focused on defining each step in the process then gathering detailed data for each step within each process. Data sheets were created and members from the Kaizen team went to gather data based upon studies that were planned or in process. Although the VSM started with the process development group there was insufficient time to gather any detail regarding this portion of the VSM. However, given the scope of the process, the timing for assessing capacity at the operational level begins after receipt of the request so the team felt comfortable proceeding without this information. The team then evaluated both the new and existing data in order to extract the VA activities, while aiming to minimize the BVA and eliminate the NVA activities.

The Kaizen team divided the small-scale manufacturing process into the following unit operations:

- 2L and 10L Prep and Autoclave;

- Production;

- Harvest Operations/Kill and Clean;

- Data Delivery.

Looking at the historical distribution of requests to the group the predominant demand was at the 2L fermentor level. The team decided that the initial focus would be limited to the 2L process stream for this initial effort. The team members then tried to identify the lean flow metrics required for the VSM assessment. The metrics shown in Table 5.1 were evaluated for each unit operation.

Table 5.1   Results of the value stream exercise

Resource Loading	Direct Labor Requirement
Cycle time	Total time required to complete all sub-tasks associated with the unit operation
Changeover time	The time required to tear down and setup for the next batch
VA time	The time spent actually creating value in the end product
BVA time	The time which is required for a process step that is out of the users control (for example, cell growth)
NVA time	The time spent which does not add value to the final end product (for example, the seven wastes)
Adjusted Process Cycle Efficiency (APCE):	Defined as the total value added time (VA)/total cycle time (C/T)—business value added cycle time (BVA): $$APCE = \frac{VA}{C/T-BVA} \times 100$$

The results of the value stream exercise are summarized in Table 5.2. This table also provides an estimate of the Process Cycle Efficiency (PCE) reflecting the process flow efficiency of the 2L Pilot operation on a stepwise basis. PCE is defined as:

$$PCE = \frac{VA}{C/T} \times 100$$

The data in Table 5.2 were used to construct the cVSM for the 2L process, and is shown below in Figure 5.1.

Table 5.2   Overall process cycle efficiency—2L operation

Unit Operation	Cycle Time (minutes)	Value Add Time (minutes)	Changeover Time (minutes)	Non-Value Add Time (minutes)	PCE (%)
2L Prep & Autoclave	735	560	223	175	76
10 L Prep	9169	332	286	197	63
Production	22666	1532	0	974	61
Harvest	438	358	198	45	89
Data Delivery	5858	117	0	5741	2
Total	**38866**	**2899**	**707**	**7132**	**7.5**

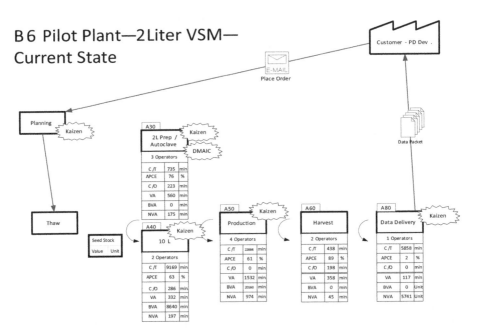

B 6 Pilot Plant—2 Liter VSM— Current State

Figure 5.1   2L Pilot Plant cVSM

Using these data the current 2L process had an estimated cycle time of 39,000 minutes or 27 days with an overall PCE of 7.5 percent. As a benchmark, a world-class PCE for batch operations of this type is considered to be ~30 percent.

Capacity improvement opportunities were divided into two categories: 1) those activities that increased product velocity or throughput through the plant, and 2) those activities that improved process quality and reliability impacting product throughput indirectly. Throughput activities typically utilize Lean tools such as 5S, Visual Factory, Rapid Changeover, Mistake Proofing and Continuous Flow techniques. Quality and reliability improvement activities typically utilize various DMAIC tools which are part of the Six Sigma framework for identifying and eliminating the root causes of the quality issues and reducing process variation.

Throughput Opportunities

Figure 5.2 shows a time series plot of throughput performance for the 2L Pilot operation over a 12-month period. This time series plot was developed using existing production data supporting the small-scale Pilot operations.

Figure 5.2 shows a 12-month trend having an average of 80 runs/month. The throughput variation computed over the same period was 24 runs, which is considered high by most references. Both the individual run and variation

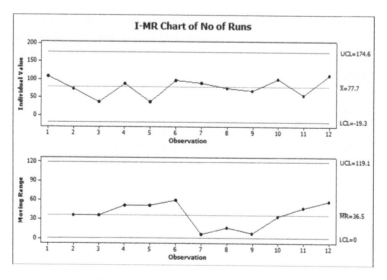

Figure 5.2   12-month throughput performance — 2L manufacturing

in individual run trends were found to be stable. A stable trend provides an indication that the 2L operation runs in a fairly consistent manner. However, given the high observed variation, the ability for the Pilot Plant operation to predict future performance was poor. Based upon the data it was possible to forecast with reasonable confidence that the 2L operation can be expected to perform about 80 runs/month on average ±16 runs/month given no outside influence or drastic changes to the overall system. The range of 16 runs is considered the 95 percent confidence interval for the average using a sample size of 12 months. Confidence intervals are used to convey the confidence in the prediction that the value of interest will fall between the limits defined. The level of confidence of the confidence interval would indicate the probability that the confidence range captures this true population parameter given a distribution of samples. It does not describe any single sample. This value is represented by a percentage, so when we say, "we are 95 percent confident that the true value of the parameter is in our confidence interval," we express that 95 percent of the observed confidence intervals will hold the true value of the parameter. Notice the 95 percent error interval for the average of ±16 runs/months is very close to the desired level of improvement per month, 18 runs. Eliminating throughput variation completely would allow the Small-scale Ops management team to easily meet half of the capacity objective of ~8–9 runs/month outright. In this sense, the month to month throughput variation was having a significant impact on the 2L operation's capacity.

The team shifted its focus to identifying the source of the variability. The team conducted a review of the batch movement control records for possible causes of throughput variation and concluded that the small-scale operations had reasonable control of their planning and scheduling activities. Using the VSM as a guide the team assembled a list of potential sources of variation to investigate:

- data handling, processing, and delivery;

- non-value add activities in the production area;

- dissolved oxygen and pH probe preparation;

- Miniferm build and preparation;

- 10L build and preparation.

Any one of the areas identified above could produce the observed throughput variation and cause processing consistency issues. The trick in realizing tangible improvement is to approach each one in an organized disciplined manner to effectively drive out process waste and time traps causing lost operational efficiency.

Assuming a 50 percent reduction in the NVA times within these processes the estimated cycle time reduction opportunities are summarized in Table 5.3 as follows.

Table 5.3 Throughput-driven estimated cycle time reduction opportunities

Process Step	Estimated Cycle Time Reduction (Days/Month)	Methodology (s)
Data handling, processing, and delivery	0.5–1	5S, VSM, Mistake Proofing
NVA activities in the production area	0.34	Rapid Changeover, 5S, Mistake Proofing
Dissolved oxygen and pH probe preparation	≤ 0.20	5S, Visual Factory, Supermarkets
Miniferm build and preparation	0.06	5S, Visual Factory, Mistake Proofing, Continuous Flow, Supermarkets
10 L build and preparation	0.07	5S, Visual Factory, Mistake Proofing, Continuous Flow, Supermarkets

The team chose to focus on the Miniferm assembly as part of the Lean DMAIC exercise and to defer the other opportunities as next steps in realizing plant capacity.

Striving for One-piece Flow

The Mini-fermentors are used in the cell culture process at the 2L batch size. The operators and technicians within the PPS group assemble the Miniferms, then sterilize them in the autoclave prior to use in production. The Miniferms are configured similarly to large-scale fermentors and require multiple connections for tubing, for monitoring, material transfer, sampling, and so on. The setup involves several dozen small components, gaskets, clamps, and so on as part of the assembly. The change parts were kept in a large cabinet and set of drawers for the operation.

The team focused its efforts on the Miniferm build and preparation area initiating a 5S and Visual Factory exercise to organize and label all elements of the assembly process. Prior to initiating the 5S exercise the team went through a short primer on achieving one-piece flow. The challenge delivered to the team was to be keenly aware of the eight wastes when contemplating how to optimize the assembly area. In addition, the team discussed Lean prinicples such as Mistake Proofing and the concept of supermarkets of reconsumable supplies as a level loading tool, which can greatly stabilize a work center within a value stream. The current area was very congested and was being used for storage of change parts for equipment, supplies, and consumables. The area contained no signage and the room housed spare parts and supplies for operations in the large-scale process and for equipment that had been decommissioned or discarded. The team removed and decommissioned all non-essential supplies and equipment in the room retaining only those supplies required to support Miniferm assembly and probe calibration. All other material was inventoried and red tagged for disposal. A general email was distributed to all potential interested departments if there was interest in any of the material. The guideline was anything that was still there at the end of the day would be discarded.

The current Miniferm assembly process utilized a team of four to six operators working on a U-shaped table layout in the center of the room. Next the team broke the assembly process down into sub-steps and was able to divide the process into two basic groups of activities that would only require two operators to complete the activities. The team identified inefficiencies in the current Miniferm assembly process specifically pertaining to ensuring all required components were available for each Miniferm assembly. To address the time lost in finding components, a detailed list of all required components was pulled together then divided by the two-step operation. These components were then bundled as kits. This not only recovered the time lost hunting down components but also ensured that there were enough components for every Miniferm assembly.

When focusing on production leveling, the team decided to group the kitted components and manage the inventory of these kits through an intermediate inventory location called a "supermarket" in Lean terms. The supermarket allowed the operation to establish a visual inventory indicator to ensure adequate component kits were always available in the assembly area where they are needed. The warehouse group was enlisted to check the inventory location twice a week. Once the inventory dropped below the green into the red indication level the supermarket was replenished with new component kits. The Miniferm assembly required that all tubing be securely fastened to the fermentor. This was one of the key contributors to contamination failures during experimentation. The group used zip ties to secure the tubing to the connectors. The zip ties were a major time issue, requiring significant dexterity to fasten them to each connector. To streamline the operation, the team built a fixture which pre-assembled the zip ties in each component kit. This allowed the zip ties to be easily and consistently attached to each connector. The team found this one step eliminated considerable manipulation time during the assembly process.

Next the team addressed the assembly layout. The Miniferms were currently being built on assembly tables arranged in a U-shape. After assembly the Minferms were then moved to a cart which was then rolled into the autoclave for sterilization. The team was challenged to think about the whole value stream from assembly through autoclaving then use, and eliminate excessive handling transfer, lifting, and so on with the ultimate goal of one-piece flow. In response the team removed the U-shaped table and purchased moveable carts that could be used right after autoclaving and component cleaning for staging the Miniferms. So instead of assembling the Miniferms on a table then transferring them to a cart then transferring them again to the autoclave cart, the team assembled them on the transfer carts then transferred them to the autoclave carts. This eliminated an entire transfer step in the process. The team briefly considered building the Miniferms directly on the autoclave carts but decided against it because the carts were expensive and could be damaged if kept in general circulation. The team will create travelers to define the order in which the Miniferms will be built, and to ensure that all operators perform the assembly task in a consistent manner. The large and small zip tie guns used to tighten the zip ties on to the tubing on the Miniferm were replaced with pneumatic guns, reducing stress and strain for the operators. Each of the pneumatic guns were hung from the ceiling over each cart and were placed in the center of the process stream for easy access.

Striving for One-piece Flow-making Believers Out of Doubters

The process flow was redesigned to support aggregated assembly tasking. In this configuration each assembly operator has a unique set of tasks at each stage in the assembly process. To ensure operator safety and minimize fatigue we recommended the operators switch assembly responsibilities after five to seven Miniferm assemblies. Switching operators as suggested has the added benefit of ensuring that all operators remain adequately trained to assemble all sections of the Miniferm while minimizing any potential chance of repetitive motion injury.

Prior to redesigning this process I suggested to the team that we apply this concept of dedicated activities with one operator setting up and the next operator finishing the operation. The team did not believe that there would be any advantage versus the current approach of both operators doing the assembly simultaneously. The new process was designed around creating Continuous Flow through the assembly process versus the current batch process. To demonstrate the advantages of the new approach a test run was conducted comparing the old and new processes. Two operators were timed in their assembly of one complete Miniferm. The old process was modified for this test in that all components were already available, rather than the usual process that requires them to navigate the room to obtain components. It required five minutes for two operators to assemble a single Miniferm using the old process.

Next the test was designed to utilize the new proposed flow developed by the team. Two operators were used, each with defined tasks. In this new configuration, the first operator would place the components on to the equipment and the second operator would chase the first operator and do the final securing of the components. The challenge was to see if there would be any difference in productivity across the same assembly period of five minutes. The results of the exercise were as follows:

Activity	Current Productivity	New Productivity
Number of Miniferms assembled	1	7

If we had fitted the components we would have expected to have seen an even higher rate of throughput with a greatly reduced likelihood of assembly error.

After assembly, all prepared Miniferms were stored on a cart until ready for transfer to the autoclaving process. It was recommended that all autoclaved Miniferms be stored in a locked cabinet along with a traveler that indicates the date of autoclaving and the unit's expiration date or "use by date." Miniferm travelers are an integral component for managing pre-assembled Miniferms in a supermarket.

The team agreed to create metrics that would track performance against the fundamental 5S metrics, and regularly review these to maintain a high level of operational efficiency.

Quality and Reliability Opportunities

A 12-month review of the 2L process quality performance was performed and revealed that success rate of completing an experimental run was about 89 percent. The variation in success rate was approximately 8 percent. An 89 percent success rate means that about 11 percent of the runs were not successful due to one of many possible quality issues. Figure 5.3 shows the Miniferm success rate trend against a lower control limit used to identify pointwise changes in time. The trend of Figure 5.3 indicates that during the 12-month evaluation period there were no significant pointwise or trendwise changes. We can interpret this plot as indicative of a process that provides a relatively consistent level of success over the period of evaluation.

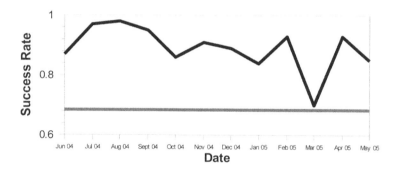

Figure 5.3   2L Miniferm processing success rate

Given an 11 percent loss rate in failed experimental runs we were interested to know if any patterns in the causes could be identified. Upon further evaluation of the quality data we identified the following breakdown rates for causes of failed lots (Table 5.4).

Table 5.4   Breakdown rates for causes of failed lots

Failure Root Cause	Incidence
Miniferm contamination	8.3%
Setup or other error	1.3%
Equipment malfunction	1.2%
Proportion of Miniferm run losses	11.0%

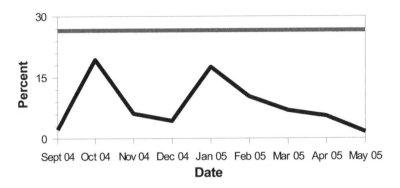

Figure 5.4   Breakdown rates for causes of failed lots

Next, the data were evaluated using Change Point analysis to see if there were any particular timing points in the trends for each of the failure causes. Figure 5.4 shows the trend of failed lots due to contamination.

While it was understood that the Pilot Plant team had been working actively to drive down failures due to Miniferm contamination, the trending analysis did not provide conclusive proof of their success. The plot in Figure 5.4 can be interpreted in two possible ways: 1) a significant change in contamination rate has not occurred, or 2) there was not enough data available to determine if a significant reduction in contamination rate had occurred. In either case, the contamination trend does not provide additional clues for improvement.

In addition to the trending and Change Point analysis of contamination failure data, an evaluation was performed to determine if any correlation existed between the number of runs processed and the incidence of contamination. The purpose of this analysis was to uncover where workload induced errors could be a possible root cause of Miniferm contamination. A significant positive relationship would be indicative of a possible workload associated cause. The results are shown in Figure 5.5.

As can be seen, there is a weak correlation between the number of runs, and the level of significance was about 5 percent. The regression plot above shows a negative relationship between the incidence of contamination and number of lots processed in a given period. While there is some weak indication of a relationship between these two factors, this relationship does not support a workload-induced hypothesis as being a cause of Miniferm contamination.

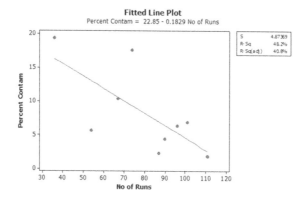

Figure 5.5 Correlation plot between Miniferm contamination incidence and number of runs

A side observation was made regarding good aseptic practice within the 2L process stream as a potential for contamination as it pertains to manipulations which take place in non-aseptic areas. One example is the manipulations that take place in the process hood that are shared with the Kill and Clean process. These shared manipulations are a potential source for contamination. We recommended that the management group commission a team to investigate ways to eliminate the potential for contaminating a batch, utilizing pipettes for aseptic operations in these non-aseptic areas.

The failure data were then evaluated to determine if there was a time relationship in the incidence of equipment setup errors resulting in failed runs. The results of this evaluation are shown in Figure 5.6.

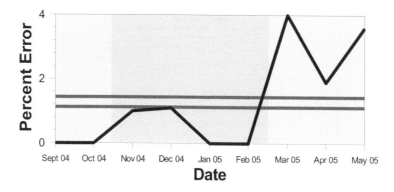

Figure 5.6 Incidence of errors/month

The data revealed there were two periods of time that indicated an increasing trend in error (see Table 5.5).

Table 5.5   Increasing trends in error

Time	Rate Change	Confidence Level
November 2004	0% to 0.53%	82%
March 2005	0.53% to 5.2%	87%

It appeared the error rate had been increasing since November of 2004. The cause of this increasing setup error rate was not well understood. We led the team in a short-loop brainstorming exercise in an attempt to uncover possible causes of this increasing error rate, but little additional causal understanding resulted from this work.

To gain further causal insight these data were further analyzed to determine if a relationship existed between the number of runs and the error rate. Again, we were attempting to connect operator work loading to the incidence of setup error within this analysis. A positive significant relationship would provide the evidence of a likely cause and promote further investigation. The results shown in Figure 5.7 provided a very weak relationship between the number of lots and error rate. The analysis inferred that workload was not a key driver for equipment setup error in this operation. We recommended that the management team undertake a Mistake Proofing analysis to try and standardize these operations with an eye toward preventing fermentor contamination.

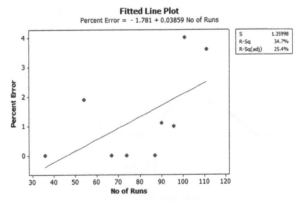

Figure 5.7   Regression plot of number of lots and error rate

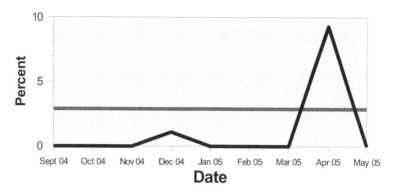

Figure 5.8   Equipment-related operational issues for the past 12 months

Finally, the data were evaluated for equipment-related issues, not associated with setup errors, over the last 12 months. The results are shown in Figure 5.8.

The data revealed no significant trends other than a point-wise change in April 2005. The data were again analyzed for workload-related issues using a regression approach. The results of this analysis are shown in Figure 5.9. As in previous analysis, we are looking for a significant positive relationship to support our hypothesis that operator workload could be a possible cause of equipment operational issues.

This analysis indicates that a significant relationship between the number of runs and equipment failure rate did not exist. It implied that operator or area workload does not appear to have a significant relationship with equipment operational issues.

The team looked deeper into the nature of the equipment failures which caused contamination failures. A review of the equipment run logs revealed that the vast majority of equipment-related issues centered on probe failures. In fact, probe reliability appeared to be a predominating factor in equipment failures over time. It was recommended that a separate Lean Kaizen be performed to address the issue of probe reliability and that the current operation remove from service any suspect probes. Additionally, it was recommended, given the high cost of each probe, that an improved probe usage tracking scheme be developed to ensure probes are removed from service for reconditioning or replacement before they fail in operation. This action would have the added benefit of greatly improving the run to run success rate, and the reliability of the overall process.

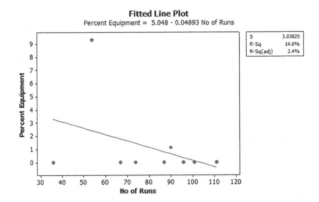

Figure 5.9   Regression plot of equipment issues and number of runs

In reviewing the quality and reliability opportunities identified in this effort, the Miniferm operation presented the greatest opportunity. In focusing upon this area, Table 5.6 presents the potential opportunities for a capacity increase, assuming a 95 percent success level/month and an average monthly load of 93 runs.

Table 5.6   Quality and reliability-driven estimated capacity increase opportunities

Process Step	Estimated Capacity Increase (Runs/Month)	Methodology (s)
Miniferm setup reduction	1–2	Mistake Proofing/Rapid Changeover
Dissolved oxygen and pH probe reliability improvement	1–2	DMAIC
Miniferm contamination reduction	7–9	DMAIC, autoclave cycle development

Long-term Capacity Increase

One side effect that was unseen by the group as a result of these intermittent equipment failures was the design creep that was cannibalizing the available fermentor capacity. Because the process development group could not afford to miss data points in their orthogonal experimental designs they over-designed the experiments, adding replicate points at each of the conditions, fundamentally doubling the size of the experiment. If the reliability of the 2L production operation could be improved through the elimination of periodic failed runs it would be possible to approach the process development group about reducing the size of their experimental design. The current designs average 18 runs in order to assure that they do not end up with missing key data. Typically, three factor designs require about nine or ten trials for a fully functional design. The current designs used by process development appear to add redundancy to minimize the chance of losing an entire design point. If the operation could demonstrate high enough reliability to match their current risk of any missing design points, ~2 percent using the present run configuration, then a potential opportunity exists to reduce the number of runs to about nine or ten, resulting in a 95–98 percent increase in the overall capacity.

Conclusion

The team was able to identify and recover the target of 20 percent capacity at the end of the five-day Kaizen. It took several weeks to develop the travelers, SOPs, and conduct training for all operators. In parallel, the team developed metrics based upon the 5S exercise and was able to maintain the area as required for the assembly operation. When the new assembly process was put into operation, the first Miniferm build was able to build 26 Miniferms using two operators in the same amount of time it used to take four to six operators to build six Miniferm units. More importantly the rate of contamination failure dropped sharply using the new procedure. Lastly the energy and enthusiasm created within the group served as a catalyst for the organization to expand its commitment to improvement using LSS tools in its large-scale operations.

Chapter 6

Applying Lean Six Sigma
on the Manufacturing Floor

There is nothing so useless as doing efficiently that which should not be done at all.

Peter F. Drucker

The most common driver for pharmaceutical companies to adopt Lean is to improve the overall manufacturing efficiency. Unlike Six Sigma, Lean Manufacturing principles do not require intensive training to adopt and to most operations and engineering personnel resonates at an intuitive level. However, for Lean to be effective in a GMP environment it is essential to frame the problem and success metrics in both business and compliance terms.

Why pursue Lean improvements? The benefits of Lean are many fold:

- faster feedback on process performance (increased learning cycles);

- improved first pass yield (improved productivity);

- improved process stability (improved throughput);

- uncovers process deficiencies (forces problem resolution);

- less work-in-process (reduced risk);

- improved customer satisfaction (flexibility and responsiveness).

One of the basic tenets of Lean is to try and understand the underlying elements that impede productivity. Lean's inventor, Taiichi Ohno, defined eight basic wastes which cannibalize the productivity of a manufacturing operation. These wastes are described as follows:

1.   overproduction

2.   excess inventory

3.   defects and rework

4.   retesting

5.   waiting

6.   excess motion

7.   transportation

8.   unnecessary efforts.

Because of these wastes, on average only about 5 percent of the overall cycle time within the value stream is actually value added as defined by the customer.

The challenge with GMP operations are there are compliance checks which are required to demonstrate the product was manufactured in a controlled manner. Optimizing processes in the absence of GMPs often results in non-compliant operations. Lean, if properly applied, however, can be a powerful stabilizing element to any process. As operators move toward standardized work practices and equipment stability improves, the true variation associated with raw materials and processing conditions become more apparent. With this increased process visibility it is possible to characterize and optimize process behavior using the tools and approaches from Six Sigma.

The Lean Kaizen

One of the most attractive elements of Lean is the ability to rapidly realize focused improvement without having to commit long-term resources which is often the case in Six Sigma projects. These rapid improvement projects are called Lean Kaizen. The Japanese word *kaizen* simply means "good change." The word refers to any improvement, one-time or continuous, large or small, in the same sense as the English word "improvement." However, given the common practice in Japan of labeling industrial or business improvement techniques with the word *kaizen* (for lack of a specific Japanese word meaning "continuous improvement" or "philosophy of improvement"), especially in the case of oft-emulated practices spearheaded by Toyota, the word *kaizen* in English is typically applied to measures for implementing continuous improvement, or even taken to mean a "Japanese philosophy" thereof.

Kaizen was made famous as a core element of TPS. In Toyota's TPS, when an abnormal event occurs the expectation is the operators and supervisors will immediately convene to analyze the event and recommend a corrective action. While a Kaizen event within the TPS usually delivers small improvements, the culture of continual aligned small improvements and standardized work practices yields large results in the form of compounded productivity improvement. This philosophy differs from the western concept of generalized improvement programs of the mid-twentieth century. Kaizen methodology includes making changes and monitoring results, then adjusting. Large-scale pre-planning and extensive project scheduling are replaced by smaller experiments, which can be rapidly adapted as new improvements are suggested.

Toyota has built a culture of improvement framed around these Kaizen events as daily process, the purpose of which goes beyond simple productivity improvement. It is also a process that, when done correctly, humanizes the workplace, eliminates overly hard work, and teaches people how to perform experiments on their work using the structured analytical methodology and how to learn to spot and eliminate waste in business processes.

In pharma we see three basic approaches to Lean Kaizen on the shop floor. The first and most common approach is to change workers' operations to make their job more productive, less tiring, more efficient, or safer. Usually buy-in is not an issue since the end-product will benefit the individual as well as the organization. These Kaizens are typically team-based and include Subject Matter Experts (SMEs), support individuals, and other team members as part of the Kaizen exercise.

The second approach is to improve equipment by designing and installing mistake-proofing devices and/or changing the machine layout. The third approach is to improve the associated procedures related to the operation. This can range from equipment setup and maintenance to troubleshooting and control approaches. All these alternatives can be combined in a broad improvement plan.

In modern usage, it is designed to address a particular issue over the course of a week and is referred to as a "Kaizen Blitz" or "Kaizen Event." These are limited in scope, and issues that arise from them are typically used in later blitzes.

The components within a Kaizen Blitz typically include:

- training members of the Kaizen team on the Lean principles that they will be applying;

- facilitating a brainstorming session to identify improvement options;

- implementing improvements by "breaking apart" the process, identifying the value added and non-value added activities and putting it back together without the waste;

- preparing an action plan which lists the activities required to complete the Kaizen process;

- identifying the measurement metrics and measurement tools to demonstrate improvement;

- soliciting active participant feedback;

- reporting Kaizen results and celebrating success.

The foundation for these activities can be found in one the oldest tenets of the pharma quality system, the Shewart Cycle, also called the PDCA Cycle. PDCA stands for Plan–Do–Check–Act. The PDCA Cycle is shown in Figure 6.1.

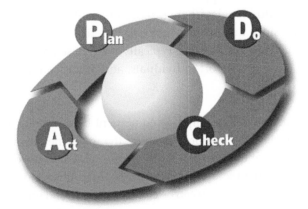

Figure 6.1   Shewart Cycle/PDCA Cycle

Applying the PDCA concept to a Kaizen Blitz, the order of activities would resemble the following:

- standardize an operation and activities;

- measure the operation (find cycle time and amount of in-process inventory);

- gauge measurements against requirements;

- innovate to meet requirements and increase productivity;

- standardize the new, improved operations;

- repeat the cycle.

These activities should be applied within the concepts of achieving one-piece flow, 5S, and standard work processes.

For Lean Six Sigma processes, the approach is to identify which portions of the business problem are potentially a result of inefficiency and which portions of the process are related to unknown process variability. However the basic tenets of PDCA are applicable to Six Sigma as well. Figure 6.2 summarizes how the tools from Lean and the tools from Six Sigma line up within the PDCA model.

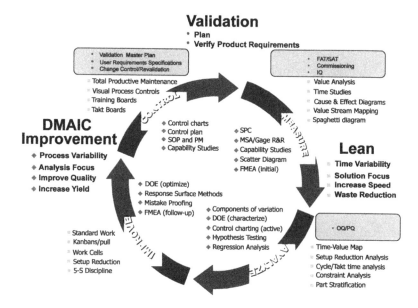

Figure 6.2   LSS and PDCA Cycle

Manufacturing Case Study

Several years ago I was involved in an optimization project with a manufacturer of transdermal systems. The product demand had exploded and despite moving to a 24/7 operation the plant was approximately 15 weeks on backorder and manufacturing was struggling to gain traction against the demand. The product was actually being manufactured for another company which had sales and marketing responsibility through a licensing agreement. Because of the escalating backorder situation, the two companies were approaching litigation and tensions were very high. The project objectives were very clear—identify existing bottlenecks and improvement opportunities that would increase productivity and alleviate the backorder situation in a timely fashion.

PROJECT CHARTER

The project was managed by a technical project manager within the OpEx group. Given the potential legal issues with the distribution partner, an oversight committee had been formed to review activities. The sponsor for the Lean Six Sigma exercise was the executive vice president of operations reporting directly to the head of the business unit. Prior to the initiation of this program there had been individual engineering and process projects focused on trying to improve productivity which had resulted in limited to no improvement. As with all LSS projects it is important to clearly define the success metrics at the outset of the improvement exercise. The team set out to establish the success metrics before beginning the improvement initiative. A simple chartering template was used to capture the project's salient information (Figure 6.3).

In addition to these metrics for success there were business constraints placed upon the team by management at the outset:

- minimal downtime would be allowed for investigative studies since the product was at 15 weeks backorder and climbing;

- at a minimum, the current approved quality standards had to be maintained;

- no more than two shifts overlap at any given time for either collecting data or deploying improvements;

- no regulatory changes could be part of the near-term solution;

- no new suppliers could be utilized for near-term solutions because of the need for raw material qualification studies and potential regulatory impact;

- the product had been on the market for over a decade so there was little to no product development information or baseline product design or meaningful characterization information available.

Within the Six Sigma world we call the success metric the "burning platform." So the burning platform for the organization was clear—nearly double the output of the operation in less than two months!

Element	Description	Team Charter				
1. Process:	The process in which opportunity exists.	Current transdermal manufacturing process can only manufacture 1.3million (MM) per week				
2. Project Description: what is the "Practical Problem"	Problem and goal statement (project's purpose)	*Increase the utilization of the manufacturing line from 50% to 65% and do sufficient process characterization studies to allow the manufacturing line speed to increase from 99 to 110 cycles/minute while maintaining a yield of >90%. Resulting in an increased revenue of $4MM/month*				
3. Objective:	What improvement is targeted and what will be the impact on Rolled Throughput Yield (RTY), Cost of Poor Quality (COPQ) and Capability index C-P, back orders, costs?	**Project Y's**	**Baseline**	**GOAL**	**Entitlement**	**units**
	The "Statistical Problem" - the measurable variable(s)	**Metric 1** utilization	50	65	-	%
		Metric 2 yield	90	90	-	%
		Metric 3 output	1.3MM	2.5MM	$50MM	units/ hr
4. Business Cases:	Expected financial improvement, or other justification.	Current backorder situation is 15 weeks. Revenue impact is estimated at $50MM				
5. Team members:	Names and roles of team members?					
6. Project Scope:	Which part of the process will be investigated and excluded.	Phased approach focusing on final Transdermal manufacturing and pouching operation				
7. Benefit to External Customers:	Who are the **final** customers, what are their key measures, and what benefits will they see?	Partner manufacturing organization that sells and markets product				

Figure 6.3   Project charter template

PROJECT STRUCTURE

The structure of the project team is a key component to how quickly and effectively it is possible to move through the LSS exercise. LSS approaches are based upon a cross-functional team make-up. The Six Sigma team and project model is based upon ensuring appropriate business and organizational alignment. The roles played by each individual are shown in Figure 6.4.

Figure 6.4   Six Sigma program and team structure

However, given the backorder situation the structure adopted for this project needed to be nimble and high visibility in order to assuage the anxieties between the two companies. A hybrid project structure was adopted which still utilized the multi-functional Black Belt core team and project champion but was combined with an external team of LSS consultants and engineers in an effort to shrink the timeline. The concept behind the hybrid team approach was to leverage internal and external resources to allow end-to-end parallel processing beginning with solution identification all the way through implementation.

By moving the design and fabrication process outside, the organization internal resources were freed up from the design and fabrication process but still available for the internal design review and change control procedures necessary to implement any improvements. The project structure is shown in Figure 6.5.

PROJECT BACKGROUND

The laminates are purchased, slit, and converted as part of the upstream process. The drug is extruded on to the laminate as part of the upstream process then the final system is assembled then held as a continuous roll prior to final processing. The equipment for the upstream processes and for the final packaging process is very old. The product is temperature sensitive and is held at 2–8°C prior to processing. The product is sold in multiple strengths and uses different laminate systems. The final processing equipment is multi-lane and singulates each patch then advances them for individual pouching. The equipment uses vision inspection to verify the lot number and expiration dating of each pouch. The equipment uses a mechanical sensor to detect empty and double-filled pouches. The heat sealing station is a 2x3 configuration mounted in a single frame operated by a single central cylinder. Historically the line was originally validated at the upper end of its run rate but over the years operators have slowed the machine down and routinely make adjustments to the pressure, speed, and temperature to achieve what is considered an acceptable seal. The pouches are burst tested to test for material defects and leak tested to test for sealing defects throughout the run. The final packaging line was run by nine operators.

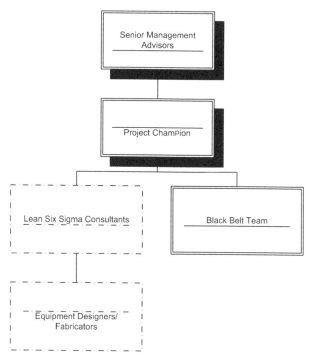

Figure 6.5   Hybrid LSS project team structure

There was a lead operator on each shift that directed the activities on the line. The operators' roles and responsibilities during manufacturing were largely assigned based upon equipment familiarity, training, and seniority. Two mechanics serviced several manufacturing areas. Each line reported to a manufacturing shift supervisor that was responsible for the manufacturing productivity and line clearance verifications in addition to Quality Assurance.

PROJECT APPROACH

The project team used a Kaizen approach in establishing its project plan. Given the rapidly deteriorating backorder situation, the team decided to break the improvement initiative into three phases. Phase 1 was essentially a Kaizen Blitz. The team would do a deep dive into the current operational data in an effort to find productivity opportunities which could move the line toward its goal of 2.5 million units a week. Phase 2 would focus on increasing the line utilization. Phase 3 would determine if the line speed could be increased without impacting the final system pouch integrity.

PHASE I—LOW HANGING FRUIT

While the objective of the LSS exercise was to improve, optimize, and stabilize the manufacturing process for the transdermal systems, there were near-term opportunities which would add to the productivity of the line quickly. A layout of the equipment, operators, and material handling equipment is shown in Figure 6.6. The organization had a fairly well developed OpEx program. Hence the team started with whatever operational data was available for the line in terms of its utilization and productivity.

The first focus was on downtime. The team did not convert the metric to Overall Equipment Effectiveness (OEE) yet, but rather attempted to understand what issues were causing the line to stop, other than setup and tear down. Every downtime event longer than one minute as logged in a log book for every lot manufactured on the equipment. Concentrating on the pouching operation, the team gathered the data and summarized it in a Pareto analysis to try and understand what was impacting the overall equipment availability. The summary data are presented in Figure 6.7.

Immediately, three key causes of equipment downtime were identified. With the current staffing the line was stopped during break periods, the passdown meeting between shifts, and during the lunch break. The cardinal rule in automation is "once it is running, don't turn it off!"

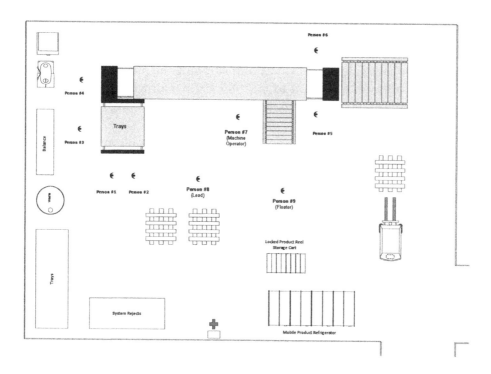

Figure 6.6   Transdermal manufacturing line area layout

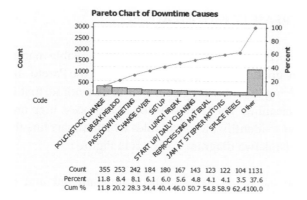

Figure 6.7   Downtime Pareto analysis

In response the team recommended the following changes to the ongoing operation: first, add two people to each shift. This would allow the operators to rotate and take rolling breaks without stopping the line. This would also allow the lead from each shift to meet for the passdown meeting without any impact to the line and allow the operators to take rolling lunch breaks.

Some may say this is anti-Lean, adding additional headcount to the line. Indeed this is exactly what the management team indicated. However, this illustrates the value of the project chartering process. The objectives were to address the project backorder situation within the compliance, regulatory, and business constraints as defined at the outset of the project. Cost was not a consideration nor was profitability. The team requested the standard cost breakdown of each strength and presentation as part of the background data gathering exercise. The cost model indicated that for these products the primary costs were in the materials not labor. In fact labor was only 6 percent of the overall standard cost, and the increase in productivity would more than warrant the additional burden. The impact of the additional headcount on all three shifts was to boost productivity from 1.3 million units per week to 1.7 million units per week, a 30 percent increase instantaneously. It is important to remember that the basic principles of manufacturing don't always require a lengthy statistical analysis. Making more, good product in the time available was the one thing we needed to work on. Adding the additional labor allowed us to boost the productivity by 30 percent without making any changes to the process.

PHASE 2—INCREASE LINE UTILIZATION

The next phase focused on trying to increase the available manufacturing time per shift to make good product. The team turned to the Pareto analysis with the intent of identifying the three biggest causes of downtime, that is, changeover, setup, start-up/cleaning. The team developed a process map and an Ishikawa diagram in order to identify those variables which impact line throughput and downtime. The Ishikawa diagram is shown in Figure 6.8.

Based upon the Ishikawa diagram the team gathered and compiled data regarding changeover and setup time by shift in an effort to better understand if there was variability that was cannibalizing line availability. The results for each are shown in Figures 6.9 and 6.10.

Based upon this data it was apparent that there is variability between the shifts which is impacting the equipment utilization. The operation had encountered some turnover and the backorder situation had pushed

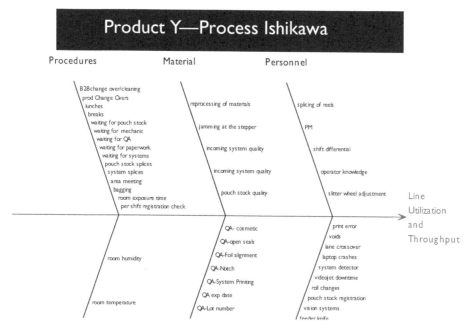

Figure 6.8   Line utilization and throughput Ishikawa diagram

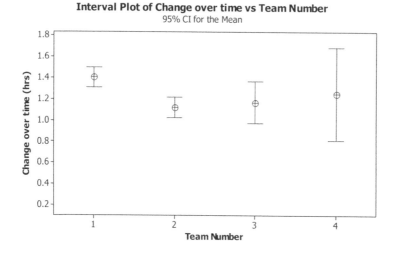

Figure 6.9   Changeover time by shift

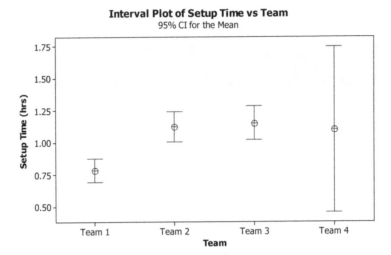

Figure 6.10   Setup time by shift

the manufacturing operation to add a fourth shift to cover the weekends. Focusing on downtime the team decided to see if the changeover and setup operations could be simplified and standardized. The team then initiated a Rapid Changeover Exercise (RCE) to try and simplify and standardize these two activities.

Rapid Changeover Exercise (RCE)

The RCE is also sometimes called Quick Changeover, Machine Changeover, Single Minute Exchange of Dies (SMED), and One Minute Exchanges of Dies (OMED). These principles are central in striving toward single-piece flow by enabling reduced lot size and hence more flexible manufacturing. The concept of rapid changeover was developed at Toyota by Shigeo Shingo. SMED and OMED are actually specific instances of rapid changeover which Shingo spent much time perfecting.

The team evaluated the data and established a target of a combined changeover and setup time of 30 minutes versus the current three-hour average combined time. The RCE team consisted of the leads from each shift along with a SME mechanic, quality, operations, and the LSS consultants. The team utilized a five-step approach to optimizing rapid changeover:

Step 1:   Document the changeover and separate events into internal or external.

Step 2:   Convert internal events to external events.

Step 3:   Streamline the internal events (simplify, reduce, eliminate).

Step 4:   Eliminate adjustments wherever possible.

Step 5:   Monitor and control.

The theoretical benefits from the methodology are illustrated in Figure 6.11.

Internal tasks were tasks that could only be completed when the equipment was down or stopped. External tasks could be performed while the equipment was running in preparation for the changeover activity and with no impact on equipment availability and productivity. The team captured every task associated with changeover and setup that required 30 seconds or more. No sophisticated software was used, rather the team captured the information on post-it notes and arranged them in an order which described how the procedure was currently performed. This by itself revealed a great deal about the shift to shift variability. Each shift lead operator had a different approach to how they handled setup and changeover. They also had their own individual tricks that

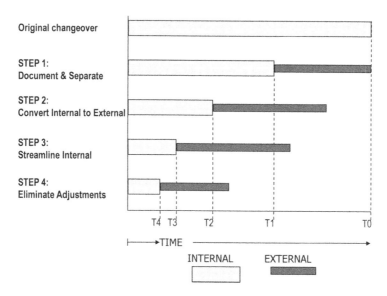

Figure 6.11   Rapid changeover methodology

they would use on their shifts for completing the tasks. Conducting an RCE can have beneficial effects beyond the data-gathering element. By creating a team of SMEs from each shift, the validity of the data is maintained as no one individual can skew the time estimates based upon their best guess. Multiple SMEs are available to validate the time required for each step. Secondarily the team gained insight into how each shift deals with challenges of the changeover, sharing best practices, and lessons learned. Finally, the culmination of the exercise clearly illustrated how much work actually goes into a changeover and setup which is both revealing and a source of pride for each shift's team.

The analysis identified several large opportunities for reducing the changeover time and setup time for the line. First the teams universally lost time searching for flashlights to look for stray patch systems that may have fallen under the machine. It was essential to make sure that all systems were accounted for before moving to another manufacturing lot. Second the team was always searching for a power pallet jack to move the remnant material back to the warehouse at the end of the run. Third the disassembly, cleaning, and reassembly of the change parts for the equipment was very time consuming because the original equipment was fabricated with viewing windows mounted by 20–30 little hex head screws. All of the screws had to be removed, cleaned, and replaced before the next lot could begin. Lastly the mechanics lost nearly 45 minutes per lot verifying the wiring to each vision inspection camera was correct. This was an artifact from an incident which happened a decade before in which the cameras had been installed incorrectly—as a result the accept/reject mechanism was not rejecting the correct systems from the line. In addition the cameras were mounted in different locations depending upon the presentation being run at the time. Mechanics had used permanent markers to provide guidance as to where the ideal location might be, although there was an adjustment step which routinely took place at the start of each run to establish the final location.

Mistake proofing (Poka-Yoke)

To address these four issues the team was challenged to identify mistake-proofing solutions which could easily be implemented in order to regain the lost time per shift. Mistake-proofing solutions have the common characteristics of being simple to design and implement. In most cases they do not cost much to implement as well. To address the flashlight issue, the mechanics hardwired lights under the equipment allowing the entire undercarriage to be inspected easily after each run. To address missing power pallet jack issues, a new pallet jack was purchased for the line and spray-painted bright pink along with the line name. The concept was to be able to quickly and easily identify if someone had borrowed the equipment for another operation. In order to address the disassembly and cleaning issue, the team had two new sets of plates fabricated for the equipment without the screws for the viewing windows. The equipment was stored on carts, one for clean components and one for dirty components. In addition, all of the mounting screws were eliminated and replaced with alignment pins and clamps. By having a clean and dirty set of plates the changeover was very fast. What had taken approximately 45 minutes to one hour to execute previously could now be done in 45 seconds to one minute.

The vision system setup issue was addressed in two steps. The wiring verification step was eliminated from the setup procedure by replacing each wire with a unique connector. The connectors precluded the mechanics from setting up the cameras incorrectly. In order to address the camera placement issue, mechanical stops were built into the new plates and clearly marked for each presentation. These mechanical stops had a simple alignment pin setup which eliminated all placement and adjustment issues with the cameras. This reduced the 45-minute setup time to less than one minute for both camera systems.

Quality checks and line clearance-standardized work flow

The last great time loss was associated with the line clearance checks between runs. The current procedure required the lead operator, the supervisor and, finally, Quality Assurance to conduct a line clearance inspection. The inspection focused on area clearance but emphasized strays and misdirected intermediates within and under the equipment. The line currently had nine operators per shift. It was perplexing how there could still be failures with so many eyes on the process. To better understand the processes the team decided to conduct a work flow analysis. Remember, the roles and responsibilities of the operators were determined based upon experience with the equipment, training, and seniority. The team analyzed the current task/work flow by operator. The results of the analysis are captured in Figure 6.12. Based upon this the team decided to try and create a future state work flow by operator. VSMs are one tool for doing this but the analysis the team was trying to perform was more of a simplification and stabilization exercise rather than an efficiency analysis. The team applied the following roadmap starting with the current state map in Figure 6.12:

Figure 6.12   Quality inspection/line clearance current work flow analysis

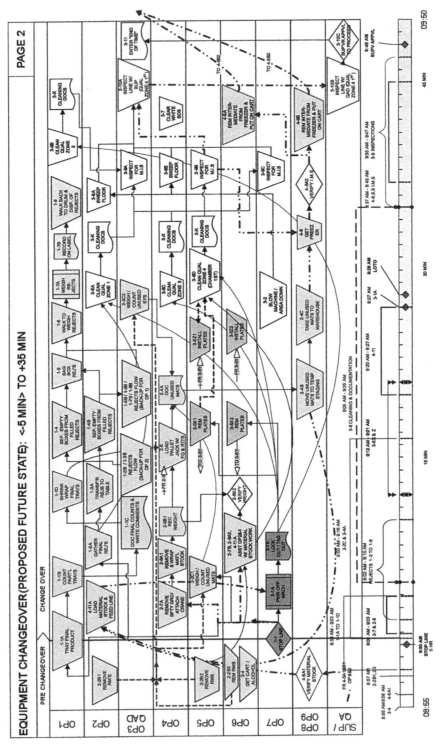

1.   assign responsibilities to each operator position;

2.   reduce the number of tasks required by the operator and create quality zones for each position;

3.   eliminate all redundant quality checks;

4.   redefine process problems associated with misdirected intermediates.

The data were captured in a swim lane diagram. The results of the analysis are shown in Figure 6.12.

The work flow analysis confirmed that, because there was no strategy to the roles and responsibilities of each operator, it was difficult to assign accountability to any one individual for any specific readiness activity. Consequently it was different every time. In the event someone was sick then the process became even more complicated as the new individual had to be limited in terms of what they could do to support the line. Following the methodology described above, the team used this current state analysis as a baseline of reassigning operator responsibilities. Each task was captured based upon technical and skill requirements as well as compliance impact. The team decided, based upon the current changeover requirements, that no operator could have more than seven specific areas of responsibilities. Simplifying and standardizing the roles and responsibilities of each operator made accountability measurable and it was much easier for each operator to become accountable to each other and to the whole process. The decision to standardize also had the additional benefit for any new operators because they now had a discrete set of tasks against which they could be trained and evaluated on reducing the overall risk and burden to the whole process. The results of the reassigned responsibilities are shown in Figure 6.13.

To try and address the redundant inspection problem, the team turned to a distributed quality model. The distributed quality model requires that multiple individuals, who are responsible for the operation of the task, are also responsible for the compliance aspect as well. This is implied within the concept of cGMPs but is not explicitly called out. In essence each individual becomes a quality designee or proxy with responsibility for the compliance aspect of the operation. The manufacturing area was divided into four distinct cleaning zones. Within each zone specific operators had responsibility for a portion of the quality zone as it pertained to line clearance. Each operator had

Figure 6.13 Quality inspection/line clearance optimized work flow analysis

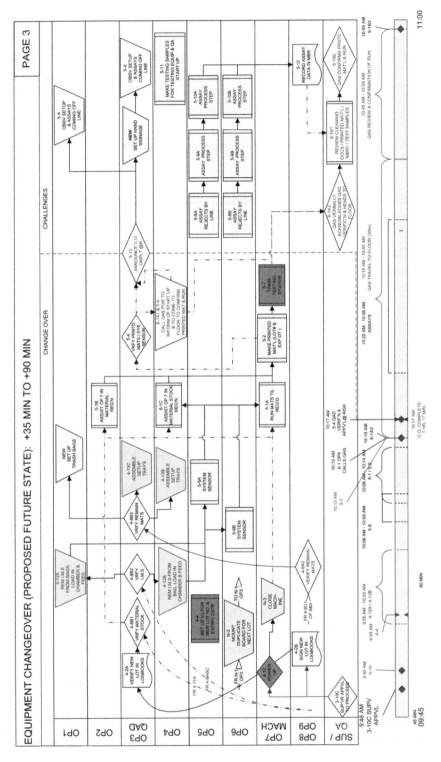

specific areas within the zone to inspect and to confirm were clean and free of product. Prior to line clearance readiness each operator, starting at the end of the equipment, would release their zone to the next operator. The team eliminated all redundant checks and performed a parallel qualification exercise with Quality Assurance under protocol to demonstrate the effectiveness of the new process. Each operator's performance was tracked and recorded until the new process had met the acceptance criteria defined in the protocol and had demonstrated capability. The four quality zones and operators impacted are shown in Figure 6.14.

The net result of Phase 2 activities was to reduce the current three-hour changeover and setup time to 56 minutes. With continued training and refinement the organization was confident that the target goal of 30 minutes was within reach. The additional two hours were added to the manufacturing time and contributed further to the overall productivity of the operation.

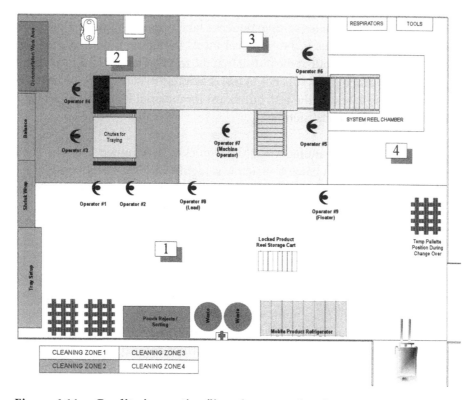

Figure 6.14   Quality inspection/line clearance cleaning zones

PHASE 3—INCREASED LINE SPEED

The last phase of the improvement initiative was to determine if the line speed could be increased from 99 to 110 cycles/minute on the final packaging equipment without adversely affecting the pouch integrity for each system. Historically the line had been qualified at the rate of 110 cycles/minute but over time the operation had slowed down the process as a result of poor deviation investigations and CAPA investigations. This evaluation would utilize Six Sigma tools rather than Lean tools in order to determine capability and impact on the final pouch and product. The focus was the heat-sealing process. The Critical Process Parameters (CPP) determined during the old validation were identified as pressure, dwell time, and temperature. In addition a strip of Teflon tape was sometimes used on the heat-sealing dies to facilitate a more consistent seal because over the years the die plate frame had begun to warp, creating an uneven interface during the sealing process. The systems are sealed using a 2x3 heat seal die array, meaning two rows of three heat seal dies each, which was held in a single frame assembly. Heat-seal quality was being established using a burst tester which inflates sample pouches taken from the line to a pressure of 25 PSI and determines if the heat seal meets the minimum burst pressure of 5 PSI. It was pointed out at the outset that the test was being used incorrectly as the purpose of a burst test is not to test seal integrity, although this is a by-product to some extent of the test, but to test for material failures which could compromise the products' stability.

A review of the manufacturing data quickly determined that the data within a batch and from batch to batch was highly variable. Before evaluating the process the team wanted to know if the burst test was capable of resolving the level of process control desired by the optimization effort.

To answer that question the team conduct a Gage Repeatability and Reproducibility (GRR) which is a structured experimental test designed to separate the contribution to variation from the operator, method, and part. GRR. A GRR of 20 percent with at least five distinct levels of categorization is typically considered the minimum level of capability of a method which will be used to steer a process. The first GRR results revealed an inordinately large operator component. The method was further defined to standardize the operator component and was able to meet the GRR metrics of <20 percent and >five distinct categories.

Given the method capability the next step was to design an experiment to determine if there was a process space which could accommodate the faster speed with no detrimental impact to the seal integrity. Operators routinely adjusted the CPPs for the process on a batch to batch basis to achieve what was perceived to be a better seal.

The team selected a Plackett–Burman design, varying the top heat seal temperature, dwell time, pressure, and speed. The experiment was blocked with and without the Teflon tape. Plackett–Burman designs have the advantage that they require less runs than a full factorial design to identify the main effects contribution but have the disadvantage that there is a confounding of interaction effects of the main variables. These designs are typically called screening designs and are used when you want to whittle down the possible set of important parameters to those whose main effects have the greatest impact on the output variable of interest. The decision to use a Plackett–Burman design rather than a full factorial design was also partially driven by the limitation that we were only given one shift in which to complete the study. Looking at historical data, and rather than risk an incomplete design, the team compromised and accepted the simpler screening design in order to complete all studies within the allotted eight-hour shift.

Figure 6.15 is a graphical representation of the regression analysis. If any combination of variables had an impact at $\alpha = 0.1$ then the histogram bar would cross the line. The model was reduced starting with CE in an effort to see if the final reduced model had any meaningful impact on the heat seal strength based upon the burst and leak test. The model was also reduced for the blocking variable, Teflon.

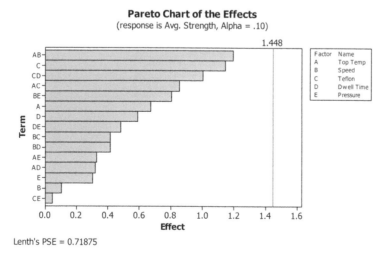

Figure 6.15   Heat sealing Pareto analysis

Based upon the analysis and the subsequent reduction, none of the CPPs or blocking variables had an effect on heat seal quality. When this model was fully reduced, the results were the identical. In a nutshell the day-to-day adjustments being made by the operators were having no tangible impact on the seal quality and the equipment could run at speeds of 110 cycles/minute with no impact of seal quality.

Conclusion

The methodical application of LSS tools by the team was able to make a tangible impact on the productivity of the manufacturing operation. With the additional capacity created by Phases 1, 2, and 3 the overall line output increased from 1.3 million units/week to 2.1 million units/week, a 61 percent increase. This equates to an annualized incremental revenue benefit of $41 million. Although this fell somewhat short of the final target for improvement, the increased capacity was sufficient to relieve the backorder situation. The improvement initiative had other tangible benefits to the manufacturing organization. Standardizing and simplifying operator responsibilities and work practices greatly aided the training of new personnel and allowed existing personnel to climb the proficiency ladder more quickly. The creation of quality zones elevated the effectiveness of line clearance, reducing the need for quality oversight and manufacturing supervisor inspection. Finally, redesigning the change parts eased stress and strain on operators and greatly reduced the need for mechanics during setup, improving equipment stability and predictability.

Chapter 7

Lean and Six Sigma as a Framework for Process Analytical Technology

The process of scientific discovery is in effect, a continual flight from wonder.

Albert Einstein

The concept of Process Analytical Technology (PAT) was formally introduced to the pharmaceutical industry in the 2004 FDA landmark guidance — *Pharmaceutical CGMPs for the 21st Century — A Risked Based Approach*. This guidance advocated a complete paradigm shift in product development — one which is based upon understanding the fundamental sources of variability in raw materials, in-process materials, and process control, rather than relying upon sampling and testing of the final product against specifications. In this document the FDA introduced the concept that traditional product release testing could be eliminated if a complete control strategy could be developed which defined and controlled the variables which had an impact on a product's performance.

The FDA followed this guidance with a PAT specific guidance issued in 2004 entitled: *PAT: A Framework for Innovative Pharmaceutical Development, Manufacturing and Quality Assurance*. At the time this guidance represented the most ambitious step in the FDA's effort to change the quality paradigm for drug development. The FDA guidance defines PAT as, "A system for designing, analyzing, and controlling manufacturing through timely measurements (that is, during processing) of critical quality and performance attributes of raw and in-process materials and processes with the goal of ensuring final product quality."

Translating this directive into action has been one of several hurdles to the broad adoption of PAT at the commercial level. There are few topics within our industry that can spark controversy as quickly as whether PAT is viable as a strategic driver for business performance in the current global pharmaceutical marketplace. One might wonder how a control solution which could have such a profound impact on business performance could be at the center of controversy regarding its usefulness? We all know change can be difficult for any industry. However, for an industry like the pharmaceutical industry, which is steeped in ritual philosophy, the concepts put forth by the FDA in these two guidances were a radical departure from what the industry has used as the basic principles for process design and Quality Assurance.

Regulatory Complexity

Part of the reluctance to aggressively pursue PAT has been the unique nature of the regulatory process for PAT implementations. To effectively apply PAT involves the application of QbD as articulated in ICH Q8(R2). Without process understanding there is no opportunity to implement closed loop control. For drug innovators, PAT represents an opportunity to maximize yield and minimize the potential for quality issues in commercial manufacturing. Together these factors should positively impact the product's true standard cost and profitability. There is no one single approach to PAT. An effective PAT implementation will require intelligently integrating mechanistic modeling with a focused QMS that controls variability at the raw material and API manufacturer through the final product performance acceptance criteria. The incentive to consider PAT or PAT approaches has increased with the issuance of the new FDA Process Validation Guidance in 2011 which is demanding a greater level of process understanding in order to satisfy the Stage 1 level of the process validation lifecycle.

Nonetheless, the regulatory process for filing a PAT process with the FDA is a significant departure from the structured process followed for a conventional NDA or 505b2 submission. In an effort to incentivize the industry to pursue PAT, the regulatory filing is flexible, meaning it can be initiated by an applicant with a scientific proposal along with the IND/NDA/ANDA. The proposal is submitted to the FDA PAT team for discussion around the appropriate level of characterization information necessary to support PAT. The members of the FDA PAT team can be found on the FDA website. The applicant has several options when pursuing a PAT solution. The applicant can choose to follow the current control strategy and a PAT team or certified PAT inspector can precede or follow the PAT implementation. The applicant can choose to file a supplemental submission such a Prior Approval Submission (PAS), a Changes-Being-Effective Supplement such as a CBE30, or choose to report the change as part of the Annual Product Report prior to implementation. This would more than likely prompt a discussion with the FDA before approval would be granted for implementation. The last scenario would be to utilize a comparability protocol submission which would contain a discussion around critical parameters, analytics, validation, timelines, and so on.

This new flexibility in regulatory filing is confusing for current regulatory groups which are used to following a strict criterion articulated by the FDA in terms of what is and is not acceptable. Depending upon the sponsors, the timing in contacting the FDA will dictate the level of risk associated with the PAT implementation.

Pharmaceutical Quality Assessment System (PQAS)

The FDA has bundled these new concepts in with a new regulatory filing process called the Pharmaceutical Quality Assessment System or PQAS. The PQAS system can be characterized as:

- knowledge-rich submissions demonstrating understanding of product and process;

- specifications based on product requirements for safety, efficacy, and stability;

- a process designed and controlled to robustly and reproducibly deliver quality product;

- regulatory flexibility based upon enhanced product and process knowledge;

- facilitated innovation and continuous improvement throughout the product lifecycle.

A comparison of a traditional NDA filing and a PQAS filing is illustrated in Table 7.1.

Table 7.1   Comparison of a traditional NDA filing and a PQAS filing

Traditional CMC Submission	PQAS CMC Submission
Quality by testing and inspection	QbD—quality assured by a well-designed product and process
Data intensive application—disjointed information without "big picture"	Knowledge-rich submission—supporting product and process design
Specifications based on process history	Specifications based on product performance requirements
"Frozen process" discouraging changes	Flexible process within design space allowing continuous improvement. For example, Stage 3 process validation
Focus on reproducibility—often avoiding or ignoring variation	Focus on robustness—understanding and controlling variation. For example, Stage 2 process validation

Before moving forward with a PAT solution it is important to have a clear understanding of which approach will be pursued from a regulatory perspective if the full benefits of the PAT implementation are to be realized.

Process Control versus Product Control

The FDA's shift from a systems-based approach to quality inspections underscored the FDA's reliance upon oversight systems rather than science to ensure product safety and efficacy. Due to the emphasis on oversight control, most firms would "lock-down" their processes and control methods once process validation was complete. Product quality was achieved through in-process offline inspection, rather than on identifying, understanding, controlling, and optimizing CPP. The FDA was not the only regulatory body to recognize this need to change how we measure and manage product quality and solicited input from its counterparts in Canada, Europe and Japan, as well as industry and academics worldwide. Several key guidance documents from The International Conference on Harmonization (ICH), ICH Q8 (Pharmaceutical Development), ICH Q9 (Quality Risk Management), and ICH Q10 (Quality Management) have become the standard for transforming organizations that aspire to the highest degree of scientific rigor in the product development process.

The FDA has fundamentally legislated these principles in the new process validation guidance in 2011. Borrowing heavily from the principles of ICH Q8, the FDA has divided process validation into three distinct stages. Stage 1 is a focused on process design. The intent is to identify Critical Material Attributes

(CMAs) and Critical Process Parameters (CPPs) which can have an impact on the product's performance. The FDA does not explicitly define what a CPP is and this has caused a great deal of confusion within the industry. I have always used the definition that a CPP is a parameter, which across its Proven Acceptable Range (PAR) has an impact of the product's final performance, for example its Critical to Quality Attributes (CQAs). In order to fully implement a PAT solution it will be essential to control all factors and CPPs which can have an impact on the product's final performance.

Lean and Six Sigma Contributions

When you think of Lean Manufacturing and Six Sigma principles it may not be immediately apparent how these frameworks can be effective in supporting a PAT initiative. Lean and Six Sigma both share the same fundamental philosophy of applying a structured process for understanding the parameters and attributes (raw material, API, and so on) that affect the process output of interest. In Lean the focus is on inefficiency driven by the eight wastes while Six Sigma focuses on the variability within the process, considering all process inputs such as raw materials, equipment, personnel, and measurement capability in an effort to steer and stabilize the process. To fully harness the potential of PAT, both Lean and Six Sigma can contribute to the final control strategy for the process. Six Sigma follows the DMAIC roadmap, however, when you evaluate the principles employed in most effective deployments of Lean improvements you will find that Lean aligns quite well with Six Sigma (see Figure 7.1).

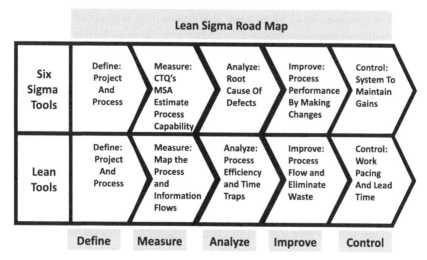

Figure 7.1   Lean Sigma roadmap

The LSS approach leverages the stabilizing impact of Lean that is achieved through standardized work as a precursor to characterizing and optimizing the process through Six Sigma. Whether an organization strategically intends to build its business strategy around PAT or intends to wade into the area as a pilot program, a structured framework built upon the principles of LSS can reduce the risk of PAT program failure.

Six Sigma and Process Analytical Technology

A criticism leveled at many PAT implementations is that efforts have focused on the application of the online analytical technology (as a replacement for offline laboratory testing), rather than on understanding control and reduction of variation. In other words, the focus has been on the technological measurement challenges rather than the improvement of product quality through process understanding. There is no question that the measurement component is a central issue to understand before pursuing a PAT technology. For example, in the early years many PAT programs attempted to measure fugacity as a surrogate for demonstrating Blend Uniformity. Defining then defending these surrogate measurements requires extensive characterization in order to demonstrate both correlation and causality with traditional process measurements.

Statistical tools for characterization and optimization of manufacturing processes have knowingly or unknowingly been a part of the pharmaceutical and biotech quality toolkit for more than 50 years. Every application of the AQL sampling plan for lot release or for a quality-driven investigation has required a baseline level of process understanding and capability in order to confidently extrapolate the results of the sample population testing to the process population. A renaissance in the more widespread use of industrial statistics by non-statisticians came with the broader adoption of LSS in the mid-1980s. A search of the PAT literature does reveal an emphasis on the use of statistical tools, particularly multivariate methods (see for example the excellent series of articles in *The PAT Journal* by Theodora Kourti, PhD on PAT and Multivariate Statistical Process Control).

The Six Sigma methodology provides an effective framework for the process characterization and optimization required by PAT that is superior to the use of statistical tools isolation. No two PAT systems are identical but adopting a methodology that clearly defines the prerequisites for process understanding can dramatically improve a PAT team's ability to stay on task. The following seven-step process is one which has proven to be effective in PAT programs and in realizing the level of process understanding necessary to implement an effective PAT control strategy.

Process Analytical Technology Case Study

Several years ago I was asked to assist a manufacturer of a controlled release tablet that had identified a number of improvement opportunities in its manufacturing process flow at one of its primary solid dosage manufacturing plants. The issue under discussion was their largest product, a controlled release tablet which utilized a high molecular weight polymer to control the diffusion of the drug during release. The drug dissolution had been variable since its market introduction two years ago, resulting in an escalating percentage of rejected lots and a higher than desirable incidence of level 2 and level 3 dissolution testing. The product demand forecast was steadily growing and the yield impact was resulting in an erosion of plant capacity, resulting in missed shipments and lost revenue to the business unit. Based upon the projected product demand and current process performance there would be insufficient capacity to meet the anticipated demand.

Analytical error had been ruled out. The lab was using the USP<711> dissolution method. A great deal of process characterization was performed as part of the investigation to identify the root cause for the variability. Based upon the product's design there were only a few mechanisms that could impact the product's dissolution profile. The primary investigation focused upon the polymer coating and the coating unit operation.

The product was a controlled release tablet which means it followed zero-order kinetics. Zero-order kinetics means the drug release rate is fixed. For example a 100mg controlled release tablet may have a drug release of 10mg/min. This is in contrast to a sustained release tablet which follows first-order kinetics for drug release. Using the same example, a 100mg sustained release tablet would have a drug release of 10 percent/min. The product dissolution curves varied significantly, however the slopes of the curves were not varying indicating the fundamental polymer diffusion characteristics were not varying. Controlled release products utilizing this simple concept of a controlled release polymer coating for drug dissolution follow a Fickian diffusion model which means, at steady state, the material flux is a function of length and time as is the diffusivity. In practical terms, this means if the porosity of the polymer coating is controlled and the thickness of the coating applied is controlled then you can expect a predictable diffusion across the membrane. Translating this mechanistically means there are only three possible ways the drug dissolution rate could change at steady state:

1.   change the surface area of the tablet;

2.   change the porosity of the polymer coating;

3.   change the thickness of the coating applied.

If one of these three mechanisms are not the root cause of the variability then, perhaps, the process is not at steady state and something else is affecting the drug diffusivity.

In evaluating the current manufacturing process, the tablet coating weight was being used as the control endpoint for the spraying process. Consequently the amount of coating material being applied was fairly consistent. An intensive investigation revealed that the coating weight was being measured and used as the endpoint for the process. Evaluating the uncoated tablet cores revealed that the weight variation and overall dimensions were not atypical. This eliminated mechanisms 1 and 3 as potential root causes. In order to determine if the process was somehow impacting the porosity of the polymer coating being applied, the uncoated tablets were tested as a baseline for comparison with different lots of coated tablets. Testing of the uncoated tablets revealed that the uncoated tablet dissolution rate themselves were varying significantly enough to warrant L2 testing per USP <711>. So it appeared mechanisms 1–3 were not the driving force for the dissolution variability.

Uncoated tablet dissolution had never been a part of the commercial in-process or release criteria so there was little to no historical data to reference. The investigation concentrated upon the uncoated tablet dissolution focusing on blending and compression unit operations as the two processes which could influence the uncoated tablet dissolution. An evaluation of the current MBR revealed that a minimum time had been specified for the blending step but no maximum time. The lubricant being used in this case was magnesium stearate which morphologically consists of particles which are comprised of stacked plates, a lot like a Pringles potato chip. As the particles are sheared during the mixing process the plates separate and distribute throughout the blend. To some extent understanding the particle size distribution of the incoming material could provide an insight. The investigation identified the root causes as periodic over-mixing of the lubricant during the final blending step, resulting in more hydrophobic surface properties. The inconsistent mixing performance was related to varying raw material properties, in particular particle size distribution, as a result of multiple lubricant suppliers. This source of variation was addressed through the establishment of a particle size distribution specification. In addition, only suppliers that demonstrated a process capability greater than 1.0 against the specification were qualified for the process.

Faced with a capacity shortfall, management decided to initiate a PAT program to determine if the inconsistent product performance issue could be addressed through the use of in-line analytical measurements and closed loop control. The decision was made to deploy a PAT team to implement improvements, with the objective of eliminating or significantly reducing the process instability.

PRODUCT PROCESS FLOW

The business unit manufactured solid dosage tablets. The process flow for manufacturing is shown in Figure 7.2. The major unit operations are compounding, granulation, milling, blending, and tableting.

Although the root cause was found to be the blending step, the tableting characterization study did identify parameters and conditions which could contribute to variability. The team decided to concentrate on the characterization and control of the blending step while applying Lean stabilization principles wherever operational variability was identified.

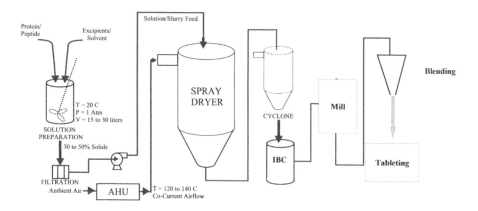

Figure 7.2   Controlled release tablet process flow

THE PROCESS ANALYTICAL TECHNOLOGY TEAM

The business unit established a PAT team which was responsible for the strategy and deployment of the project. A project team was established that consisted of experts from across the business unit, our consulting firm, and an automation supplier. Although they were not a formal LSS organization, the pharmaceutical company had experience using many of the LSS statistical tools. The PAT team decided to apply the Six Sigma structure to the project from the outset because of the perceived advantage of a toll-gate approach with its built in checks and balances. In addition, the Six Sigma approach afforded the team the ability to clearly articulate the success metrics for the individual stages of the project, as well as for the project as a whole, and align the project with current business objectives and strategy from the outset. The toll-gate approach and the use of incremental success metrics were instrumental in garnering senior management support from across the organization throughout the project. The team's first task was defining the deliverables and ensuring consistency with current business objectives. As described earlier, the plant was suffering from a capacity shortfall. The team met and summarized the situation as follows:

1.   At the current manufacturing rate the inconsistent tablet dissolution profile was costing the company $32MM in scrap on an annual basis not including the cost of poor quality associated with handling rejected material.

2.   The rejected lots accounted for 18 percent of the plants overall capacity.

3.   The estimated cost to the business unit because of the dissolution failures or Stage 2 testing requirements was approximately 6 percent of the standing work-in-process cost.

Based upon this assessment the team determined it was appropriate to proceed with the PAT project. It is important to note that the objective in this case, from a business perspective, was not to replace the quality overhead associated with the tablet release, but rather to prevent the loss of product and associated capacity cannibalization due to poor dissolution. This greatly simplified the initial regulatory strategy for the project while leaving the door open for filing to replace product release testing with an in-process control strategy, at a future date.

THE PROCESS ANALYTICAL TECHNOLOGY MODEL

In evaluating the application of the Six Sigma to the project there were several models the team explored. The classic Six Sigma DMAIC model provides a good framework for objective scientific inquiry and is typically used to improve existing processes (and products). However, the team felt that Design for Lean Six Sigma (DFLSS), with its focus on the development of new products and processes, could provide a better technical and quality mindset for this PAT project. DFLSS models provide a structured, phased approach to the design of a product, process, or service with Six Sigma and efficiency as key design criteria. Today, it is essential to incorporate some sort of risk management component to the product and process analysis exercise. Risk management was easily incorporated in the approach, as FMEA is a standard DFLSS tool, although simpler risk matrices are acceptable as well. Integration of DFLSS with PAT answers the criticism of some current PAT implementations as focusing too much on online analytical instrumentation as opposed to understanding, controlling, and reducing the sources of process and product variation. This DFLSS toll-gate approach to PAT provides the additional advantage of providing a set of success criteria for completion of key milestones within each phase of the process so the "gate" can be closed. Comparison of progress with such criteria provides objective evidence of incremental team success (that can be celebrated and communicated to the rest of the organization) and helps prevent team self-delusion that can occur in other improvement models.

The DFLSS models under consideration were Define, Measure, Analyze, Design, Verify (DMADV), Identify, Design, Optimize, Verify (IDOV), and Define, Characterize, Optimize, Verify (DCOV) (see Figure 7.3).

SIX SIGMA PROCESS ANALYTICAL TECHNOLOGY

We decided to use the DCOV DFLSS model, with its focus on process characterization and optimization as a basic framework. The basic model was modified to create more discrete toll gates within each major phase of the framework. The following were the final toll gates established for the project:

1. **Identify**: Clearly identify key elements of the project, including: regulatory strategy, regulatory commitment to identifying a defensible control strategy, and CPPs. How do these CPPs relate to the products CQAs? Also capture the key considerations from development documentation and historical process improvement activities.

2.   **Stabilize and Characterize**: Standardize all equipment setup and changeover procedures. Mistake-proof where possible. What are the CPPS that have been characterized as they relate to the in-process control strategy?

3.   **Define**: What is the defined design space for the process?

4.   **Optimize**: What is the control space that defines the allowable Normal Operating Range (NOR) in order to maintain the process within the design space?

5.   **Measure:** What analytical solutions are possible surrogates for the existing offline measurement systems?

6.   **Automate:** What control solutions can be applied to leveraged?

7.   **Verify:** Prepare a proof-of-concept, process model.

8.   **Validate**: Complete the equipment qualifcation, method validation, and comparability study.

DMADV	IDOV	DCOV
1. Define – What is the new process, product or service? Why is it needed?	1. Identify – What are the needs and requirements of the customer?	1. Design– What are the needs and requirements of the customer?
2. Measure – What are the customer requirements? How do we translate these into design requirements?	2. Design– How do we translate customer needs and requirements into a product design?	2. Characterize – How do we translate customer needs and requirements into a product design? What are the key process input variables (KPIV) that affect customer requirements (KPOV)?
3. Analyze – What are the design alternatives? How do we select the best design concept (High Level Design)?	3. Optimize– How can we optimize the design to minimize variability and meet customer requirements?	3. Optimize– How can we optimize the design (KPIV) to minimize variability and meet customer requirements (KPOV)?
4. Design– What is the design realization (Detailed Design)?	4. Verify – How do we verify the design meets customer requirements?	4. Verify – How do we verify the design meets customer requirements?
5. Verify – How do we verify the design meets customer requirements?		

Figure 7.3   Lean DFSS models

The PAT model adopted is shown in Figure 7.4.

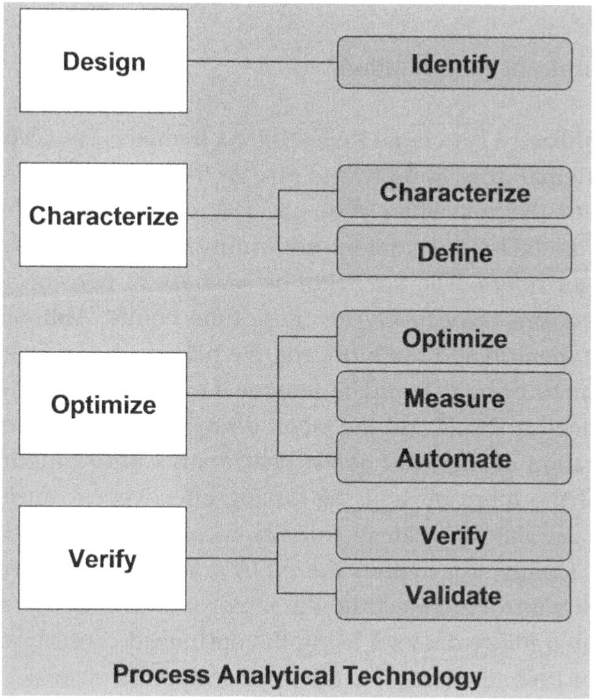

Figure 7.4   Design for LSS model applied to PAT

Within each of the phases there are a set of deliverables that must be completed to ensure all project requirements are met. Each will be discussed as follows:

Identify

In the identify phase the PAT team is tasked with identifying the design criteria for moving forward with the PAT strategy. On the process side the team developed a process map in order to identify the Process Input Variables and the Process Output Variables the team would focus on to design the PAT solution. The input variables for the blending process identified were as follows:

- granulation particle size distribution;

- mixing time;

- intensifier arm;

- lubricant particle size distribution;

- lubricant concentration.

The process utilized a Patterson–Kelley 100 cu. ft. mixer. The CMC commitment during the original drug filing was to mix for 5–15 minutes, at a mixing speed of 10rpm with the intensifier arm on. The measurement for this process step filed in the NDA was content uniformity and tablet dissolution at two, four, and eight hours. The specification was 10–20 percent, 21–60 percent, and 61–100 percent respectively for these time points. Although the process step required them to add the lubricant, the product had a low percentage of API, so content uniformity could be an issue if under mixed or over mixed. The team determined it would use the tablet dissolution, API content uniformity, and concentration of lubricant as the benchmarks for evaluating the content uniformity of the lubricant and the mixing effectiveness during the mixing process. The regulatory strategy initially focused on establishing a control range which ideally was within the NDA commitment. Since PAT focuses on a feedback control architecture, the intent was to establish a scientifically rigorous comparability data set using the optimized control range then steer the metrics for PAT automation to the same endpoints.

At this and subsequent phases, success metrics were established for the phase. Progress and metrics were presented to management at a toll-gate review meeting with management giving the team approval to move to the next phase or to take additional action to resolve any open issues. For any open issues the team would then submit a formal corrective action in order to get management's approval to then move to the next phase. This process continued through the subsequent six phases.

Stabilize and characterize

Before conducting any characterization studies, the team conducted a 5S and rapid changeover exercise for the blending and compression operations. The focus of the effort was to try to minimize and stabilize the contribution to variation from the equipment which could confound any characterization study. The compression operation was the primary focus. Clean and dirty carts were fabricated for change parts along with dedicated change parts. For the blending study, the equipment considerations were minimal but the sequence and procedures used to load the blender varied. A retrospective review of the process development data indicated that there was no evaluation of the impact of granulation particle size distribution or lubricant particle size distribution. Lubricant concentration was evaluated along with mixing time. Neither evaluation used an orthogonal experimental design, hence the data could not be regressed. The development data evaluated 1 percent and 2 percent lubricant concentrations. Based upon this development work the key CPP identified was mixing time with the product CQA being tablet appearance and dissolution. Tablet appearance was representative of the tablet compression process. A final concentration of 1.5 percent was chosen.

Given the lack of information from the original development work a characterization study was initiated to evaluate the impact of lubricant concentration, mixing time, and whether the intensifier arm was used. The ICH Q8 guidance describes this evaluation as defining the knowledge space for each process. The mean granulation size and particle size distribution and lubricant particle size distribution data were measured and kept constant for the study. The results indicated that lubricant concentration and mixing time were both CPPs at all three dissolution points. The intensifier bar did not have an effect. Key product CQAs measured were drug dissolution, drug content uniformity, and tablet appearance.

Define

ICH Q8 discusses identifying the optimum design space for the process. The design space is a sub-set of the overall knowledge space for the manufacturing process. A graphical representation of the relationship between the knowledge, design, and control space we discussed earlier is shown in Figure 7.5. In evaluating the influence of key process inputs the team focused upon a tiered approach to reducing PAT risk. It was agreed the minimum acceptance criteria was to achieve drug content uniformity. Once the control space was established for this then the behavior of the lubricant would be evaluated. The objective was to find a control space in which drug content uniformity was guaranteed along with lubricant content uniformity.

The knowledge space defines the total process space within which the process inputs can be varied. Some parameters at their limits may not produce acceptable product and some parameters may have no impact on the critical product CQAs of content uniformity, dissolution, and lubricant content. The intent of this exercise is to determine the CPPs and their PAR and NOR range limits. The design space represents the range within which acceptable product can be manufactured. This used to be called the validated range in the 1987

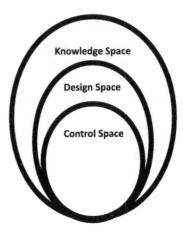

Figure 7.5   Process understanding through QbD

process validation guidance. It focuses upon those parameters which impact product CQAs. The control space represents the range within the design space in which acceptable product can be guaranteed to meet product specifications.

The PAT team initiated a follow-on study, designed to characterize the design space. The lubricant concentration was fixed at 1.25 percent, and the intensifier bar was not used. In order to evaluate the impact of granulation particle size distribution the percentage fines were evaluated. Two different suppliers of lubricant were also evaluated. All satisfied the revised specification for the lubricant. The DOE evaluated the following possible input parameters:

- Mixing time: 6–14 minutes;

- Granulation percent fines: 11–36 percent;

- Lubricant lots: 1–2.

The study revealed that mixing time was a CPP and granulation particle size distribution was a CMA for drug dissolution at the two, four, and eight-hour time points. The lubricant lots were not significant contributors to either drug content uniformity or lubricant content uniformity. All lots passed level 1 dissolution testing as defined by USP<711>.

Optimize

The next step in identifying the final processing space is to identify the control space. The control space represents a range of critical parameters within which the process will yield an output within the specified range of the products CQAs. It also represents the basis for the control architecture to be adapted for the PAT solution. Powder mixing theory states that there are three components that dominate Blend Uniformity: granulation/blend particle size, powder bulk density, and Van Der Waals forces. Of these, granulation particle size is the most significant factor. Given that granulation particle size distribution was identified as a CPP, an investigation was initiated to understand the variability in the final granulation and milling steps up stream of the blend step. Modifications to the milling setup to control the feed rate of granulation were made. A screening study was repeated to determine if the new granulation particle size distribution was still a significant contributor to Blend Uniformity and it did not come up as significant using a confidence interval of 95 percent. Based upon this, the control strategy focused upon establishing a baseline of lubricant distribution which could serve as the comparability criteria for the PAT solution downstream. The team did not focus on the drug content uniformity since the knowledge and design space studies had moved the process away from the edge of failure while characterizing the variability around the CPPs that could affect drug content uniformity.

Measure

The challenge in developing an in-line metric for ensuring proper mixing of the lubricant was that there was no offline test currently being performed for lubricant content in tablets. A baseline examination of tablets manufactured during the control phase was performed using Mass Spectroscopy (MS) in order to understand the variability around the control space. Tablets which exhibited poor dissolution were also evaluated from the original failed lots. The results of the MS data are shown in Table 7.2. The most striking observation was that the tablets which exhibited poor dissolution had significantly higher levels of lubricant present.

Table 7.2   **Lot to lot tablet lubricant content**

Tablet Lot	Percent Lubricant	Comment
Lot A	1.58	Control Space Lot
Lot B	1.68	Control Space Lot
Lot C	1.77	Control Space Lot
Lot D	1.54	Control Space Lot
Lot E	1.60	Control Space Lot
Lot F	1.44	Control Space Lot
Lot G	1.46	Control Space Lot
Lot H	1.63	Control Space Lot
Failed Lot 1	4.62	Failed S1 Dissolution
Failed Lot 2	6.48	Failed S2 Dissolution
Failed Lot 3	3.58	Failed S1 Dissolution
Failed Lot 4	5.44	Failed S2 Dissolution

Automate

The team had sufficient understanding of the behavior of the current process, its CPPs, and product CQAs to move to identifying an automation solution. The PAT team included an external automation firm which possessed a strong understanding of process, Design for Six Sigma and GAMP5 to complement their experience in custom automation. This is a significant consideration given the intimate relationship between the technical solution and quality and regulatory considerations for the project. Having a solutions provider that possesses the systems to integrate the requirements of QbD is a significant advantage in developing the scientific argument that the in-line solution is a surrogate to the offline analytical solution and for generating the necessary documentation trail to support validation downstream in the project.

The team approached the automation solution in phases. The first phase was designed to ensure there was solid understanding of the existing process performance using offline analytical tools. Tablet performance was currently measured using High-performance Liquid Chromatography (HPLC) for content uniformity and potency assessment and tablet dissolution coupled with UV spectroscopy to determine the tablet's release profile. Currently no offline assessment was performed on the lubricant, which had been found to be a key factor in achieving the desired tablet dissolution profile. The second phase was to establish a correlation with the new surrogate analytical method. The last phase was to demonstrate that the hardware solution and control algorithm resulted in tablets that satisfied the product's release criteria.

The team focused upon establishing a correlation between an offline and in-line method for lubricant concentration. Tablets were analyzed using MS and FT-NIR in an attempt to establish the correlation. The results are shown in Table 7.3.

Table 7.3   Analytical method comparison

Tablet Lot	MS% Lubricant	FT-NIR% Lubricant	Observation
Lot A	1.57	1.46	Control Space Lot
Lot B	1.68	1.55	Control Space Lot
Lot C	1.61	1.52	Control Space Lot
Lot D	1.72	1.63	Control Space Lot
Lot E	1.60	1.55	Control Space Lot
Lot F	1.56	1.50	Control Space Lot
Lot G	1.51	1.38	Control Space Lot
Lot H	1.50	1.47	Control Space Lot
Failed Lot 1	5.41	5.11	Failed L1 Dissolution
Failed Lot 2	7.05	6.89	Failed L2 Dissolution
Failed Lot 3	4.12	4.03	Failed L1 Dissolution
Failed Lot 4	3.82	3.77	Failed L2 Dissolution

A correlation curve was developed and based upon this data and the team felt it had a viable method to proceed with developing an in-line solution.

Verify

The verify step is used to establish a basic proof of concept that the principles of the solution are viable. In the previous phase the team attempted to establish a correlation between the offline and in-line measurement systems. The adherence to the Six Sigma methodology had narrowed the control space such that the process should have been sufficiently far from the edge of failure. The first focus was establishing a measurement for the lubricant in the blending step which could be used to dictate the blending time. Since the confidence was high that material mixed for 7–12 minutes resulted in tablets which had acceptable dissolution then a correlation with lubricant concentration could be one trigger used to prevent over mixing the blend. With the selection of FT-NIR as the measurement tool a blender was modified with a self-contained analytical probe and analyzer, equipped with a wireless transmission system as a proof of concept system. The final control space screening study was repeated. Measurements were taken from the in-line sensor and tablets were tested using MS for lubricant content. In addition, tablets were tested for content uniformity, potency, and dissolution at the three, four, and eight-hour time points.

All dissolution timepoints had a process capability greater than 1.33 (Four Sigma process). The process capability chart for the eight-hour timepoint is shown in Figure 7.6:

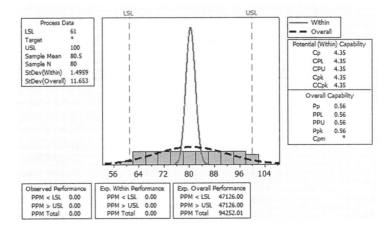

Figure 7.6   Process capability curve for eight-hour tablet dissolution timepoint

The results illustrate that the FT-NIR system was capable of controlling the process and delivering compliant product. Based upon these studies a change control notice was initiated and the production equipment was modified.

Validate

The final step in the process was to validate the equipment and process. The Six Sigma process dictated the elements to be completed as follows:

- generate the final development report;

- baseline the equipment;

- modify operational SOPs;

- modify maintenance SOPs;

- modify calibration program;

- software validation—Part 11 compliance;

- IQ/OQ/PQ;

- MS and FT-NIR method validation;

- regulatory update.

Conclusion

The statistical tools and toll-gate process of Six Sigma provides a best practice process for characterizing, controlling, and reducing process variation that is necessary to successfully deploy PAT. The partnership of Lean Six Sigma and PAT was intended to characterize and implement a control and measurement solution which would minimize the likelihood of a controlled release tablet failing dissolution. The team used a modified DCOV model sub-divided into eight phases designed to ensure that the basic requirements of the ICH Q8 requirement for QbD are satisfied. The framework ensured all aspects of the project were addressed in an efficient and methodical manner and that the scientific rigor necessary to implement an in-line control architecture was integrated throughout the process.

Index

Printed and bound by CPI Group (UK) Ltd, Croydon, CR0 4YY

18/10/2024

01776204-0013